VOLUME 470 NOVEMBER 1983

THE ANNALS

of The American Academy *of* Political
and Social Science

ROBOTICS: FUTURE FACTORIES, FUTURE WORKERS

Special Editor of this Volume

ROBERT J. MILLER
Department of Anthropology
University of Wisconsin
Madison

SAGE PUBLICATIONS *BEVERLY HILLS* *LONDON* *NEW DELHI*

THE ANNALS

ERICA GINSBURG, *Assistant Editor*

Editorial Office: 3937 Chestnut Street, Philadelphia, Pennsylvania 19104.

For information about membership (Individuals only) and subscriptions (institutions), address:*

SAGE PUBLICATIONS, INC.
275 South Beverly Drive
Beverly Hills, CA 90212 USA

From India and South Asia, write to:

SAGE PUBLICATIONS INDIA Pvt. Ltd.
P.O. Box 3605
New Delhi 110 024
INDIA

From the UK, Europe, the Middle East and Africa, write to:

SAGE PUBICATIONS LTD
28 Banner Street
London EC1Y 8QE
ENGLAND

**Please note that members of The Academy receive THE ANNALS with their membership.*

Library of Congress Catalog Card Number 83-050917
International Standard Serial Number ISSN 0002-7162
International Standard Book Number ISBN 0-8039-2170-5 (Vol. 470, 1983 paper)
International Standard Book Number ISBN 0-8039-2169-1 (Vol. 470, 1983 cloth)
Manufactured in the United States of America. First printing, November 1983.

The articles appearing in THE ANNALS are indexed in *Book Review Index; Public Affairs Information Service Bulletin; Social Sciences Index; Monthly Periodical Index; Current Contents; Behavioral, Social Management Sciences;* and *Combined Retrospective Index Sets.* They are also abstracted and indexed in *ABC Pol Sci, Historical Abstracts, Human Resources Abstracts, Social Sciences Citation Index, United States Political Science Documents, Social Work Research & Abstracts, Peace Research Reviews, Sage Urban Studies Abstracts, International Political Science Abstracts,* and/or *America: History and Life.*

Information about membership rates, institutional subscriptions, and back issue prices may be found on the facing page.

Advertising. Current rates and specifications may be obtained by writing to THE ANNALS Advertising and Promotion Manager at the Beverly Hills office (address above).

Claims. Claims for undelivered copies must be made no later than three months following month of publication. The publisher will supply missing copies when losses have been sustained in transit and when the reserve stock will permit.

Change of Address. Six weeks' advance notice must be given when notifying of change of address to insure proper identification. Please specify name of journal. Send change of address to: THE ANNALS, c/o Sage Publications, Inc., 275 South Beverly Drive, Beverly Hills, CA 90212.

Origin and Purpose. The Academy was organized December 14, 1889, to promote the progress of political and social science, especially through publications and meetings. The Academy does not take sides in controverted questions, but seeks to gather and present reliable information to assist the public in forming an intelligent and accurate judgment.

Meetings. The Academy holds an annual meeting in the spring extending over two days.

Publications. THE ANNALS is the bimonthly publication of The Academy. Each issue contains articles on some prominent social or political problem, written at the invitation of the editors. Also, monographs are published from time to time, numbers of which are distributed to pertinent professional organizations. These volumes constitute important reference works on the topics with which they deal, and they are extensively cited by authorities through-out the United States and abroad. The papers presented at the meetings of The Academy are included in THE ANNALS.

Membership. Each member of The Academy receives THE ANNALS and may attend the meetings of The Academy. Membership is open only to individuals. Annual dues: $26.00 for the regular paperbound edition (clothbound, $39.00). Add $6.00 per year for membership outside the U.S.A. Members may also purchase single issues of THE ANNALS for $6.95 each (clothbound, $10.00).

Subscriptions. THE ANNALS is published six times annually—in January, March, May, July, September, and November. Institutions may subscribe to THE ANNALS at the annual rate: $45.00 (clothbound, $60.00). Add $6.00 per year for subscriptions outside the U.S.A. Institutional rates for single issues: $10.00 each (clothbound, $15.00).

Second class postage paid at Beverly Hills, California.

Single issues of THE ANNALS may be obtained by individuals who are not members of The Academy for $7.95 each (clothbound, $15.00). Single issues of THE ANNALS have proven to be excellent supplementary texts for classroom use. Direct inquiries regarding adoptions to THE ANNALS c/o Sage Publications (address below).

All correspondence concerning membership in The Academy, dues renewals, inquiries about membership status, and/or purchase of single issues of THE ANNALS should be sent to THE ANNALS c/o Sage Publications, Inc., 275 South Beverly Drive, Beverly Hills, CA 90212. *Please note that orders under $20 must be prepaid.* Sage affiliates in London and India will assist institutional subscribers abroad with regard to orders, claims, and inquiries for both subscriptions and single issues.

THE ANNALS

of The American Academy *of* Political *and* Social Science

RICHARD D. LAMBERT, *Editor*
ALAN W. HESTON, *Associate Editor*

FORTHCOMING

PAYING FOR CULTURE
Special Editor: Patricia A. McFate
Volume 471 — January 1984

POLLING AND THE DEMOCRATIC CONSENSUS
Special Editor: L. John Martin
Volume 472 — March 1984

THE FUTURE OF AMERICAN UNIONISM
Special Editor: Louis A. Ferman
Volume 473 — May 1984

See page 3 for information on Academy membership and purchase of single volumes of **The Annals**.

CONTENTS

BOOK DEPARTMENT CONTENTS

INTERNATIONAL RELATIONS AND POLITICS

AFRICA, ASIA, AND LATIN AMERICA

EUROPE

UNITED STATES

SOCIOLOGY

ECONOMICS

PREFACE

Kindergarten, Crestwood School, Madison, Wisconsin:

Andrew . . . Tara . . . and Tim . . . knelt before a corner table, concentrating on the game board before them. Tim reached out and spun an arrow in the center of the board. It whirled, then pointed at one of several robot parts illustrated. "I got a leg with roller skates," he exclaimed, eagerly adding the part to his own composite picture of a robot.

What is a robot? "I don't know; we haven't learned that yet," said Andrew. "I think it's full of mechanical parts," offered Tara.[1]

We are all still learning what a robot really is. This volume is an attempt to place the robot in both technological and social contexts. Today the primary application of robots is in manufacturing, and so the volume emphasizes the connection between robotics and the workplace. In one of its projected developments, the factory itself becomes a practically autonomous machine—a robot—requiring a minimum of human attendance and producing a constant stream of goods.

Is this factory of the future not in some ways a realization of the story of "the mill that grinds salt"? Will our enthusiasm for the creation of the ultimate machine be as disastrous as the fable implies? In the folk tale the machine goes out of control. There are many warnings, some apparent in this volume, that the applications of the robot and its nonmobile counterpart, the computer, already pose problems for which we have only partial answers.

The drive to complete automation is most evident in the context of the factory and the office. There, at worst, humans are gradually replaced by smart machines, the pace of work is set by the machines, and even formerly skilled jobs will be filled by machines. The implications of such developments are not lost on workers in those industries, such as communications, where the pace of automation preceded current developments in robotics.

Our volume is selective and problem setting. Limited by space, we cannot examine robotics in military applications, in the home, or in agriculture. While each article deals with what exists, all probe the major questions that foreseeable robotics applications will bring into the public policy arena. Where technology is limited today, research under way suggests improvements within a two to five-year span. Where only a few completely automated factories are operative today, the model and experience gained from operation is built into the goals of the advocates of the high-technology society. To attain the full information society in which some estimate that the working population will be 90 percent in nonmanufacturing industries, roboticization must be part of the scenario of development. As an evolving system the robotic-information society depends on how rapidly the capabilities of the technology evolve. Our first section, concentrating on the technological aspects of robotics, emphasizes how far we have yet to go in making the completely automated factory the norm in a new system of production.

However, if a new system is implicit in robotics, old systems, like old people, cannot simply be brushed aside. Many social problems of robotics will stem from the need to introduce robotics into existing systems. Though completely new factories, designed as robotic environments from the beginning, may be built, older plants must be remodeled or become hybrid human-robotic units, mixing old and new machines, old and new workers. In the excitement generated by technical change in the factory, it is easy to overlook the ramifications of the change in workplace, work habits, and expectations of management. Half of this volume is devoted to setting out some dimensions of the problems to be faced in employment-related education, factory and community connections, public policy, and the valuation of human abilities.

1. Wisconsin State Journal, 4 Apr. 1983, sec. 4, p. 1.

We begin with a view of the Japanese experience, for surely this home of robotics and a working factory of the future has some significance as we in the United States rush to catch up. We explore the current state of the art in robotics technology and its uses in the proposed flexible manufacturing systems. Developing such factories poses problems to designers and industrial engineers, and leads us into issues confronting educators. We can only touch on those aspects of education immediately affected, such as managerial education, vocational-technical colleges and their expanding roles in both community-service and intermediate-educational levels, and engineering colleges. All the articles on education emphasize a common point: there must be major changes in perspective and delivery of education relevant to the robotic-computer-based society.

Finally, before the fully automated factory becomes the norm the future workers will still include many who are present-day workers. We stress here three categories, chosen to stand for the workers most immediately facing the future as three related technologies converge on them: industrial robotics, office automation, and telecommunications. In each of these contexts the workers who must be retrained, eliminated, or never hired, and the unions that represent them or may face reduction of membership, are confronting the issues being debated in educational circles.

To complete the outlines of the evolving system—having begun at the environmental and international levels—it is only appropriate to look at the relationship between the factory and the community, and between human and robot as individual actors shaping the total system of the future.

The series of articles is organized so that one may in fact read from front to back or from the end to the beginning. Read forward or backward, the total volume says much the same as Daddy Warbucks said in a recent *Orphan Annie* sequence. Warbucks is planning to automate a mine. His workers want to know what he is going to do with the robot he has brought; they ask him whom they are up against. Daddy says, "I'm a pussycat compared to your real enemy—one you can neither fight nor reason with. . . . It's not a 'who.' What you're up against is the Future."

But even Daddy is unaware of the realities. He is still dealing with the old image of the robot—semihuman form, arms and legs and TV vision, voice reception and production, a complex microprocessor brain programmed with artificial intelligence. And that image of the far-future robot can only serve to divert the workers, the managers, and the public of today from the real future that is being shaped by today's robots and computers.

ROBERT J. MILLER

NOTE: I wish to express here my gratitude to all the contributors, for their enthusiastic cooperation in the conceptualization and completion of this issue of *The Annals*. Each author's work speaks for itself, but it should be noted that all have cut into extremely busy schedules to submit their contributions. Space limitations and the design of the volume has prohibited some authors from expanding on their subjects to the extent that they might have done otherwise. Many are currently writing articles, books, and research reports probing the edges of the problems outlined here. If each had been allowed unlimited space and unlimited time, what a series of definitive volumes could have emerged!

One author, whose work is not included here, must be mentioned. Mr. Will Todd, a graduate of the Wisconsin industrial relations program, had prepared a bibliography of works—primarily articles—appearing in professional journals listed in various business sources. Space again forced us to drop this contribution. The field, as Mr. Todd said in his introduction, has been expanding so rapidly that one cannot list adequately the numbers of new journals and magazines, popular articles, and infrequent cross-disciplinary studies appearing in the last two years. Rather than cut the original articles appearing in the volume, the special editor reluctantly chose to drop the review.

ANNALS, *AAPSS*, **470**, November 1983

The Human: Alien in the Robotic Environment?

By ROBERT J. MILLER

ABSTRACT: The development of robotics, combined with computers, poses a problem even more critical in some respects than the loss of jobs. A new system is evolving, transforming the workplace environment but with ramifications beyond the workplace. Design of environments suited to the logic of robotics, given smart machines, need not consider humans as critical elements in the system. Modification of human environments raises questions of human adaptability and the nature of the man-machine relationship.

Robert J. Miller is professor of anthropology at the University of Wisconsin. He is a member of the board of directors, Council for the Understanding of Technology in Human Affairs, Inc., has participated in various professional conferences on technology and social change, and has a technical background in electronics. He is a contributing editor to the Journal of Cultural Futures Research, *writing the column "TechnoTopics." His current interests are robotics, computers, communications, and space.*

In the ideal goal of the mobile household servant Robot, sensors will be needed to compensate for the imperfect mechanical and physical world of the human environment.[1]

A man, viewed as a behaving system, is quite simple. The apparent complexity of his behavior over time is largely a reflection of the complexity of the environment in which he finds himself.[2]

The *Robotics Age* editor and Herbert Simon both pose the problem of behavior in relation to environment. Simon develops his argument to suggest an approach and a method to utilize "computer simulation as a tool for achieving a deeper understanding of human behavior."[3] Paradoxically as we design machines that can simulate, emulate, and go beyond human capabilities in particular environments, we begin to confront intellectually uncomfortable questions about the adaptability of humans.

In one view robotics is simply a more challenging phase of automation. Automation is the application of machines to tasks that are repetitive, standardized, and established for, but not by, the machine. Automation is old: the Jacquard loom, the Swiss music box and the conveyor belt are simple devices that can be placed in the history of automation. But an earlier meaning of "automation"—found in definitions of "automaton," "automatic," and so on[4]—suggests that the popular image of robots, as self-acting machines that can establish tasks for themselves, has merit.

Engineers fought a losing battle to avoid using the term robot. Until a few years ago they recognized that the term carried the earlier image derived from literature, movies, and television. They were careful to use terms such as "manipulator" and "numerically controlled device" to convey more precisely the function of the machine in question. This attempt was undermined by the development of new technologies, by the popularization of the computer as a thinking machine, and by the renewed emphasis in movies on machines that could emulate humans in all respects except two: self-replication and emotion.

Perhaps the designers of the so-called manipulators simply responded to a changed conceptual environment, in which the replication of human behavioral functions was no longer viewed as an attempt to create a Frankenstein monster. After all, even humans are today being reconstructed and enhanced with mechanical-electronic parts, computer designers now explore the combination of organic and electronic elements, and the ability to build more behaviorally flexible machines is developing almost exponentially.

However, exaggerated notions of what robots are, are becoming, or can do still fill the movies and popular magazines. It is possible to convince many that robots—meaning any machine that is flexible, reprogrammable, and does work—are about to make humans obsolete in most industrial, and in many white-collar contexts. Recent projections for the future revive old concerns and old dreams. The old fear rises: will the machine take over from

1. Editorial, *Robotics Age,* May-June 1983, p. 9.

2. Herbert Simon, *The Sciences of the Artificial* (Cambridge, MA: MIT Press, 1969), p. 25.

3. Ibid., p. 22.

4. *Oxford Dictionary of the English Language,* 1971 ed., p. 144; cf. also *American Heritage Dictionary of the English Language,* 1969 ed., p. 90.

man? Are we building our own replacements? Are the mad scientists and technologists building self-acting machines that may over time diminish the value of human-ness? The old dreams also rise: machines will completely release humans from physical and hazardous labor, will free humans to pursue intellectual—more properly human?—activities, and will produce sufficient quantities of goods that only the problem of distribution need concern us.

The realities, as our volume demonstrates, are still far from either the fears or the dreams. Robotics, even in its industrial applications, has many technological steps to make before complete replication of human biomechanical capabilities—especially in thought and adaptability to new environments—is possible. Yet in a more sober vein some of the fears are well founded. In some industrial and office settings, human labor is being reevaluated, and the nature of the human-machine combination is being redefined.[5]

While attention focuses upon the extension of human capabilities to machines, and on the displacement or deskilling of humans,[6] a possibly more significant aspect of robotics is implicit in the development of the factory of the future. When we turn from the immediate emphasis on the machines to the system of humans and machines in a workplace, we recognize that the robot—in fully developed form—emulates humans in an unexpected way. As there was a logic of human development of systems to control and make a new environment suited to humans, so is there a logic of robotic automation.

The logic of robotic automation requires the creation of an environment in which the robot may function most effectively. The mistakes of the early literary and experimental roboticists—attempts to build controllable but self-acting machines out of human parts, early mechanical robots simulating human forms but with limited functions—lay in the assumption that the robot had to function in the "imperfect mechanical and physical world of human environments." In the past the complexity of behavior demanded of the robot has been limited by the need to suit the existing imperfect workplaces that were developed as a combination of human-machine environments.

With technological advances robotic behavior now allows for more human-like complexity of behavior; but the behaviors need not be combined in one package. The contemporary roboticist, unless attempting to design the perfect home robot, can look forward to building environments suited to the self-acting machine. The factory of the future is such an environment; it is in fact itself a massive robot. Those humans who may still be needed to attend the robot, or to perform tasks still too complex for the robot, enter the world of the robot and must adapt to it.

The behaviors demanded of the human, if they are to reflect the complexity

5. See Gene Bylinsky, "Those Smart Young Robots on the Production Line," *Fortune,* 17 Dec. 1979, p. 96: "on weekends the two robots work alone, an employee coming by every twelve hours to load their parts and feed the watchdog."

6. "Past applications of technology in the work place, as well as present evidence, suggest that future technologies will further simplify and routinize work tasks and reduce opportunities for worker individuality and judgement." Quoted from an interview with R. Rumberger and H. Levin, Stanford Institute for Research on Educational Finance and Governance, *Info World,* 5(20): 32 (16 May 1983).

of the robot environment, are not at all clear. Skills developed for use with so-called dumb machines gradually become superfluous as the machine becomes more complex in behavior. Timing of human actions may need to become more precise or at least be determined by the pace of the machine. If the human is, presumably, controlling a number of machines, his or her attention ideally must be concentrated and unflagging; consider images of the attendants of roboticized nuclear plants, or the actions of flight controllers at airports. As the capability of the machine expands, enhanced by more complex computer programs, the logic of the new environment becomes apparent: as soon as possible, eliminate the human factor.

In a strange way we are forced by robotics to look again at the human history. The logic of robotics is, of course, the same logic by which humans have built their own environments with the aid of machines. Machines enabled humans to redesign the natural environment by eliminating impediments to the evolution of human-machine symbiote. It is to the alteration of the environment that we must look for the threat of robotics, not to the development of a total machine to replace us—the classical robot. The new machines—the robot and its own symbiote, the computer—requires alteration of existing environments if they are to realize fully the complexity of behavior of which they are becoming capable. In the process of deploying robots in the service of humans, the imperfections of the human environment become impediments to their further development. Thus the environment, not only the robot, at some point must be changed, just as in the human past the human-animal-nature environment had to be changed. Imperceptibly in some cases, dramatically in others such as the factory, the self-acting machine with or without human form confronts us with the challenge of a new kind of human adaptation.

If the logic of robotics has validity humans will also begin to alter their environments to utilize available robots best. As the robots become more versatile, the design of living habitats will begin to elicit behaviors appropriate to the robotically efficient environment. The visions of the totally automated home, workplace, city, transportation systems, entertainment centers, and so on have not yet been abandoned!

I am not anthropomorphizing robots; I do not assume that "they" will demand such changes. It is people who have developed and deployed robots in the latest phase of automation. The goals of people, perceiving the potential for complex machine behavior, dictate the eventual automation of all repetitive, monotonous, hazardous, mechanical tasks in human environments other than that of the factory. As humans earlier altered their environments with machines, the surroundings housing those machines—homes, schools, offices, cities—all took on particular forms in which the machines could operate or be operated most effectively, if not most efficiently. And there was reciprocity, of a sort: machines were and are designed to be more user-friendly, to fit into the existing or slightly modified human environment more easily.

It is also argued frequently that modern human social organization reflects the environments created by the process of mechanization and industrialization. Robotics is only a part of the new technological environment being created by our smart machines. The human ability to design this total environment such that it remains a human environment—one allowing for human imperfections—requires perception of the totally new man-machine system now evolving. The robot and the computer force reevaluation of the ways in which humans have thought of and used the machine. Similarly robotics and computerization raise new questions about human behavior and adaptability. Both robots and computers have been derived from human models; both can in some ways go beyond the capabilities of the human.

Almost 20 years ago Norbert Weiner framed the issues we really confront today:

> A goal seeking mechanism will not necessarily seek our goals unless we design it for that purpose, and in that designing we must foresee all steps of the process for which it is designed, instead of exercising a tentative foresight which goes up to a certain point, and can be continued from that point on as new difficulties arise. The penalties for errors of foresight, great as they are now, will be enormously increased as automatization comes into its full use.[7]

7. *God and Golem, Inc.* (Cambridge, MA: MIT Press, 1964), p. 63.

ANNALS, *AAPSS*, **470**, November 1983

Japanese Robotics: Challenge and—Limited—Exemplar

By LEONARD LYNN

ABSTRACT: At the end of 1982 Japan's robot population was triple that of the United States, and Japan's robot producers seemed positioned to dominate this strategic new industry. Underlying this early success was an elaborate set of government policies including low-interest loans, special depreciation allowances, and government-coordinated research projects. Many argue that the Japanese have been successful not only in developing their robotics industry but also in coping with the social impacts of this new technology. Their success, however, should not be overstated. Some Japanese fear that the introduction of robots may lead to record unemployment by the end of the decade. Women and certain other groups may already be disproportionately bearing the costs of the change in technology. This suggests that while we may have much to learn from Japan, it is a mistake to regard Japan as a paragon. What is needed is for Americans to study the Japanese experience with robotics, just as the Japanese routinely study our experiences, not to borrow uncritically but to seek new ideas and insights.

Leonard Lynn is an assistant professor of social science at Carnegie-Mellon University in Pittsburgh. He has a Ph.D. in sociology from the University of Michigan and degrees in East Asian studies from the University of Oregon. Lynn lived in Japan for a total of eight years, most recently as a Fulbright fellow. He is the author of How Japan Innovates: A Comparison with the U.S. in the Case of Oxygen Steelmaking *(1982) and several articles on Japanese technology.*

NEW industries often carry the stamp of certain entrepreneurs. We think of Henry Ford and automobiles, for example, or Thomas Watson and computers. These men did not invent the products with which they are associated, but they found new and better ways to produce, use, and market them. To an important extent they built the automobile and computer industries. Ford and Watson were Americans, and America long dominated the automobile and computer industries.

An article in a highly respected Japanese opinion journal suggests that a Japanese will be the Henry Ford of the robotics industry.[1] Perhaps he will be Seiemon Inaba, the president of Fanuc; indeed the stream of impressed foreign visitors to Fanuc's so-called unmanned robot production plant at the foot of Mount Fuji seems reminiscent of past generations of foreign visitors to Henry Ford's Model T plant at River Rouge. Another contender is Toshio Kono, president and founder of Dainichi Kiko. Kono started his company in 1971 when he was 30 years old. A decade later leading firms from around the world were vying for the right to sell his robots.[2]

The U.S. dominance of the automobile and computer industries contributed to U.S. success in a host of other industries. Similarly dominance in industrial robotics may well influence performance in numerous other industries. Thus we should be concerned about the fate of the U.S. robotics industry and its ability to meet this challenge from Japan.

1. Yukio Shimura, "Robotto sangyo saizensen," *Chuo Koron Quarterly, Management Problems,* Spring 1982, pp. 52-63.

2. "Dainichi Whiz-Kid Aims to Build a Better Robot," *Japan Times Weekly,* 31 July 1982.

There is a second challenge from Japanese robotics—the competitive strength of Japanese industries that heavily use robots. At the end of 1982 about 22,000 programmable robots were being used in Japan. The United States, with twice Japan's population, had fewer than one-third as many robots—about 6800. The United States also had fewer robots per capita than Sweden and West Germany.[3]

The U.S. lag is disturbing because the use of industrial robots gives clear competitive advantages. It adds design versatility and reduces materials waste and labor costs. A critical problem is whether or not the United States can meet the challenge of a Japanese industry that intensively uses robots. And if we race to match the Japanese in using robots, we may face another problem. How do we cope with the resulting changes in the employment structure? We need to know how effectively the Japanese have managed this transition and what we can learn from their experience.

THE INTRODUCTION OF ROBOTICS IN JAPAN

In 1954 George C. Devol, Jr. patented what was called universal automation or

3. The estimate of the number of Japanese robots is restricted to machines that are robots by the U.S. definition. The estimate was made by Paul Aron, executive vice-president of Daiwa Securities, and the author of two widely cited reports on the Japanese robotics industry. Other estimates based on the less restrictive criteria range up to 100,000. Some sources say the Japanese have about 70 percent of the world's industrial robots. See for example, Kazumasa Iinuma, "Robotto kenkyu no sai zensen," *Zokan Chuo Koron—Robotto Shakai Nippon,* 25 Oct. 1982. The estimated U.S. robot population is from H. Allan Hunt and Timothy L. Hunt, *Robotics: Human Resource Implications for Michigan* (Kalamazoo, MI: W.E. Upjohn Institute for Employment Research, 1982), p. 8.

unimation. This formed the basis for industrial robots, and the leading robot maker in this country is Unimation, which started out by using Devol's patents. A prototype of the Unimate was made in 1960, and in 1962 a Unimate was sold for commercial use to General Motors. Another American firm, AMF Thermatool, also introduced an industrial robot in 1960. This was the Versatran, for "versatile transfer."

It was AMF Thermatool that introduced robotics technology to Japan. In May 1966 the company showed a film of the Versatran robot to executives of a trading company in Tokyo. The trading company in turn showed the film at various Japanese firms. A Versatran was brought to Japan in 1967 and demonstrated. At the end of the year the first industrial robot to be installed for commercial use went into operation at Toyota.[4]

A year later, in 1968, license rights to the other important American robot, the Unimate, was assigned to Kawasaki Heavy Industries. Kawasaki soon became Japan's leading producer of robots, and it is claimed in the Western trade press that two other major producers, Fujitsu Fanuc and Mitsubishi, based their early robot designs on the Unimate.[5]

The Japanese, then, trailed the United States in both the production and the installation of industrial robots. As we have seen they now enjoy a substantial lead in both respects. Why were they so eager to use this new technology? And how did they manage to build up their capability so quickly?

WHY AND HOW THE JAPANESE MOVED SO FAST WITH ROBOTICS

Some of the popular press in Japan asserts that the Japanese have an unusual affinity for robots, or more precisely that they lack a mysterious allergy to robots that appears to afflict Westerners. The blurb promoting an article on robots in one of the mass circulation weeklies is typical: "In contrast to Europeans and Americans, who regard robots with hostility and repulsion, Japanese dote on robots as though robots were their offspring."[6]

There is some speculation in the Japanese press about the reasons for the mysterious Western allergy to robots, and efforts have been made to trace its origins in religion or culture. These are often fascinating, but I find a more mundane explanation of the Japanese enthusiasm somewhat more plausible.

In the late 1960s there was increasing concern in Japan about a growing labor shortage. A Ministry of Labor study projected that if Japan's then current 9 percent annual economic growth rate were to continue, there would be a need for 1.4 percent annual growth in the labor supply between 1970 and 1977. The growth in the labor supply, however, was expected to be only around 0.7 percent. A deficit of some 3.1 million

4. Seiichi Suzuki, "Robotto ga hajimete Nippon ni kita toki: American AMF-sha Basatran no omoide," *Robotto,* 7: 96-97 (no date).

5. Robert Ayres, Leonard Lynn, and Steven Miller, "Technology Transfer in Robotics between the U.S. and Japan," in *U.S.-Japan Technological Exchange Symposium* (Washington, DC: Japan-America Society, 1981), pp. 77-115.

6. "Robotto sangyo wa Nippon keizai no 'kyuseishu' ni nareru ka," *Shukan Asahi,* 12 Nov. 1982, p. 175; my translation.

workers was expected by 1977.[7] Thus it is not surprising that the theme of a symposium on robotics in Japan in 1967 was, "What can the robot do for a society that is short of labor?"

The Japanese have also long been concerned about another problem, the division of their economy into an elite sector of large internationally competitive firms and a relatively backward sector of smaller firms. Productivity, working conditions, and wages in the smaller firms have lagged far behind those in the major firms. The smaller firms, however, have been very important to the Japanese economy. They frequently serve as subcontractors for the larger firms, providing important flexibility to the system. Thus the Japanese government has long tried to help increase the productivity of the smaller firms.[8] The impending labor shortage threatened to exacerbate the small firm problem because, it was believed, these firms would be unable to compete with larger firms in getting an adequate supply of skilled labor.

Once there was a consensus that robots could solve the projected labor problem, Japanese policymakers started working to promote this new industry. The government took both direct and indirect steps to improve the level of Japanese robotics technology and to facilitate the widespread use of industrial robots in Japanese industry. Many of these steps sought specifically to encourage smaller firms to introduce robots.

In 1971 the Industrial Robot Roundtable was established. This was the first trade association of its kind in the world, predating the Robot Institute of America, for example, by three years. In 1972 the roundtable was reorganized as the Japan Industrial Robot Association (JIRA). For a year JIRA was a *jigyo dantai,* a type of public corporation intended to promote economic or social policies. Then it became a *shadan hojin,* an incorporated private association under the Ministry of International Trade and Industry (MITI). JIRA receives subsidies from the machinery industry promotion fund, part of which comes from automobile racing profits.[9] Its executive director is a former MITI official.

JIRA continues to be a very active super trade association. It publishes journals, makes promotional films, organizes symposia, collects and translates foreign technical literature and patents, and conducts surveys on the use and projected use of robots in Japan. Some of its activities go far beyond those of an American trade association. In 1981, for example, it channeled some $3 million in interest-free loans from the machinery industry promotion fund to 13 firms in the robotics industry. JIRA is instrumental as well in promoting leasing arrangements and low-interest loans for robot users.[10]

7. Walter Galenson with Konosuke Odaka, "The Japanese Labor Market," in *Asia's New Giant,* ed. Hugh Patrick and Henry Rosovsky (Washington, DC: Brookings Institution, 1976), pp. 587-671.

8. The standard Western source on the problem of Japan's smaller firms is Seymour Broadbridge, *Industrial Dualism in Japan* (Chicago: Aldine Publishing Company, 1966).

9. Kanji Yonemoto, "Sangyoyo robotto no genjo to shorai no mitoshi," *Robotto,* 7: 3-7 (no date). See also Kuni Sadamoto, ed., *Robots in the Japanese Economy* (Tokyo: Survey Japan, 1981), p. 67.

10. JIRA, "Showa 56-nendo jigyo hokoku," *Robotto,* Aug. 1982, pp. 4-8.

The Japanese government has also directly implemented a number of important measures to encourage the use of industrial robots. Several were included in the 1978 Law for Extraordinary Measures for Specific Machinery and Information Industries. Firms installing robots from 1980 through 1983 were given the regular depreciation allowance plus an additional 13 percent for the first year of installation. Below-market interest loans were provided for the purchase of robots. Small firms can get special loans from the Small Business Finance Corporation to install robots that will enhance workplace safety. In addition to these efforts by the national government, local governments have also set up loan and leasing programs to help small enterprises in their regions install robots as a part of modernization.[11]

Another very important measure was the establishment of the Japan Robot Leasing Company Limited (JAROL) in April 1980. JAROL was organized under the auspices of MITI. Its initial capital came from 24 robot manufacturers and 10 insurance companies, but it is backed by a loan syndicate that includes the governmental Japan Development Bank and Long-Term Credit Bank, as well as the Industrial Bank of Japan. JAROL offers leasing terms to users that are far more attractive than those offered by existing commercial leasing firms; it also provides systems engineering and consulting services.[12] In its first two years of operation it leased some 500 robots to around 200 different companies, mostly smaller firms.

Aside from ensuring that a large number of firms would use robots—and thus stimulating a high level of demand for robots—the government also took measures to encourage the development of robotics technology. By the late 1970s a dozen government laboratories were engaged in research related to robotics. Some of this activity was under the umbrella Large-Scale Research Projects coordinated by MITI's Agency for Industrial Science and Technology. These are projects designed to develop technology seen as being in the Japanese national interest, but which may involve too much risk for an individual firm to undertake. One of these projects sought to build an unmanned machine plant by 1980. Such firms as Hitachi and Kawasaki Heavy Industries, JIRA, and the government Mechanical Engineering Laboratory were involved. The project was unsuccessful in its larger goals and was suspended, though it seems likely that much of the technology developed under it helped strengthen the Japanese robotics industry. Another project was initiated in 1977 with the aim of developing a Flexible Manufacturing System Complex with Lasers. By 1983 the project was intended to make Japan the world leader in systems for small batch production. In 1983 still another robotics-related project was planned, a seven-year project to concentrate on the development of intelligent robots.[13]

11. The 1978 law is translated and discussed by Julian Gresser in U.S., Congress, House, Committee of Ways and Means, Subcommittee of Trade, "High Technology and Japanese Industrial Policy: A Strategy for U.S. Policymakers," 96th Cong., 2d sess., 1980. Specific measures are discussed in various issues of the JIRA publication *Robotto.*

12. "Nihon Robotto Risu Kabushiki Kaisha no seiritsu," *Robotto,* June 1980, pp. 4-7.

13. *Robot News International,* June 1982.

WHERE ROBOTS ARE BEING USED

More than half the robots in the world are now being used in the automobile industry, and Japanese automakers have been prominent in the race to introduce robots. Nissan reportedly uses robots to carry out 90 percent of the welding at its Zama plant. It plans to use some 200 robots to handle some 95 percent of the welding at the new truck plant it is building in Smyrna, Tennessee. Most of these robots are being imported from Japan. Robots installed by Japanese firms operating in the United States may be another factor that will put pressure on potential U.S. users.

Toyota has 80 robots on the welding line of its Tahara 2 Works, which replaced some 50 workers; one person is able to supervise all the robots. Honda has installed two robot painting lines at its Sayama Works that supposedly have advanced the automation of painting to new levels. According to a recent report in the *Japanese Economic Journal,* human workers are about to disappear at this plant. It is not clear, however, that the Japanese automakers are generally ahead of the Americans in their use of robots. Some stories that have appeared in the press, such as a *Washington Post* story reporting that 67-worker shifts produced some 1300 cars a day at Nissan's Zama plant, have turned out to be wildly exaggerated. The Zama plant actually employs not 67 people to a shift, but 2000.[14] U.S. managers who visit the Japanese plants are typically disappointed in what they see. Indeed it is difficult to attribute the recent competitiveness of the Japanese auto industry to its use of robots. In the late 1970s such Japanese automakers as Honda became very highly competitive before they started using robots on a significant scale. This is not to say, however, that the growing use of industrial robots by Japanese automakers will not increase their advantages over the U.S. industry. It surely will if we do not keep pace with them.

It is important to note that the use of robots is quickly spreading beyond the auto industry in Japan. In fact since 1980 the Japanese electric appliance industry has been a bigger customer for industrial robots than the auto industry, and a recent survey suggests that by the end of the decade large numbers of industrial robots will be used in hospitals, the service industries, agriculture, and fisheries.[15]

THE JAPANESE ROBOT INDUSTRY

The Japanese robotics industry differs from the American in two respects that may have broad implications for the future. First, the Japanese industry is far more diverse in the number and types of firms in it. Second, far more of the Japanese firms started out by making robots for their own use.

One reason for the greater diversity of the Japanese industry is simply that it has far more firms than the American. Virtually all of the American robot

14. This myth was debunked by Harley Shaiken on the MacNeil-Leher Report, "Robotics," broadcast 29 July 1980.

15. In 1980 36 percent of robot sales in Japan were accounted for by the electric machine industry, compared to 30 percent by the automobile industry and 10 percent in the plastic molding industry. JIRA, *Sangyoyo robotto choki juyo yosoku hokokusho,* Mar. 1981. See also Kensuke Shima, "Hiseizogyo bunya de no sangyoyo robotto choki juyo yosoku," *Robotto,* 34: 4-9 (Apr. 1982).

production is accounted for by half a dozen firms. Unimation, the largest producer, has about 40 percent of the American market. Cincinnati Milacron accounts for another 30 percent. The Japanese industry has about 150 firms, though the products of some unknown number of these firms would not qualify as robots by the American definition. The largest Japanese producer, Kawasaki Heavy Industries (KHI), accounts for only about 8 percent of the Japanese market. Moreover KHI leads only in the production of spot welding robots. Yaskawa Electric leads in arc welding robots, Kobe Steel in spray painting robots, Fanuc in assembly robots, and Aida Engineering in press robots.[16]

The range in types of firms in the Japanese robotics industry is extremely broad. Some are highly specialized. Aida Engineering, for example, specializes in the production of hydraulic presses; it is a rather small firm with fewer than 1000 employees. Smaller still is Dainichi Kiko, the specialized robot producer mentioned at the beginning of this article as possibly providing a basis for Toshio Kono's claim to be the Henry Ford of robotics. Danichi Kiko was established with a modest $5000 in capital. It still has fewer than 200 employees, compared to the 25,000 of KHI. Indeed KHI produces not only robots but also railroad rolling stock, ships, and motorcycles. KHI is not alone among the major Japanese firms in this industry, nor is it the biggest. Others include Hitachi, Matsushita, and Mitsubishi Electric—all of which are several times the size of KHI. The largest American firm that is a major producer of industrial robots is Cincinnati Milacron, with sales of not quite $1 billion in 1981; firms such as Hitachi and Matsushita had sales of $15 or $16 billion in 1981. Indeed there are about 15 Japanese firms larger than Milacron producing robots.

When large firms become involved in robotics, things can quickly begin to happen on a grand scale. Hitachi, for example, reportedly assigned some 500 researchers at its laboratories to carry out robotics-related research. Matsushita entered the industry later than some other firms but is reportedly planning to increase its robot production to 1000 per month by the end of the 1980s.[17]

Will size carry the day in robotics? The issue is by no means clear. Smaller entrepreneurial firms were primarily responsible for the advance of the semiconductor industry in the United States.[18] Sharp managed to hold both the Japanese and international calculator markets against competition from such behemoths as Hitachi and Toshiba, and Fujitsu has outpaced these larger firms in the computer industry. Given the diversity of the Japanese firms in the robotics industry, however, it would appear that whether it is the multi-industry giant or the small entrepreneurial firm that is most suited to dominate in robotics, Japan will have a number of firms competing among themselves for preeminence.

What of such American giants as General Motors, IBM, General Electric,

16. *Japanese Robotics: The Takeoff* (Tokyo: Bache Halsey Stuart Shields (Japan) Ltd., 1982).

17. "Matsushita: Seeking Industrial Markets While Staying Strong in Home Products," *Business Week,* 27 Sept. 1982, pp. 72-80.

18. This was not true in Japan. See John Tilton, *International Diffusion of Technology: The Case of Semiconductors* (Washington, DC: Brookings Institution, 1971).

and Westinghouse? These firms have all been interested in robotics and have all carried out their own research. But in 1981 and 1982 these and other firms were rushing to import technology from the Japanese. General Motors linked up with Fanuc, IBM with Sankyo Seiki, General Electric with Hitachi, and Westinghouse with Komatsu and Mitsubishi.[19] This may be a market-holding strategy, but there are few instances in which American business has successfully dislodged the Japanese once they have become established in an industry. Indeed there are claims that Japanese robots offer some of the same competitive strengths associated with other Japanese products. The executive director of JIRA is quoted as saying, "When the same robot is made in the U.S. and Japan, the Japanese has far higher quality." He compares the Unimate made by Unimation unfavorably with the Unimate made by KHI.[20] The president of Unimation, Joseph Engelberger, said in an interview that because KHI receives orders in a much larger volume than Unimation, it can supply robots for 60 percent of the cost of the U.S.-made robots.[21]

Another important difference between the U.S. and Japanese robotics industries is that while major American producers such as Unimation and Cincinnati Milacron have concentrated on general purpose robots, many of the Japanese have developed special purpose robots, often for their own use. Fanuc attempted to combine a general purpose robot with machine tools, found it desirable to custom design a robot for use in its manufacturing cells, and thus entered the robotics industry. Companies such as KHI are so diversified that they can try out robots in-house in a wide variety of applications. A number of other Japanese firms have developed special robots to meet their own needs: Seiko now sells robots it developed for the assembly of watches, Pentel sells robots it developed to assemble felt tip pens, Yamaha has made a large number of robots for in-house use that it will soon place on the market. And so it goes with such firms as Nippon Electric Corporation, Fujitsu, Toshiba, and Toyota.

THE SOCIAL IMPACT OF ROBOTICS IN JAPAN

The Japanese are now using more than three times as many robots as we are. This lead is even more impressive if we recall that Japan has only about half the U.S. population. Thus they have moved a little further into the robotics age than we have. What can we learn from the Japanese experience?

A common view is that the introduction of robots has gone very smoothly in Japan. The typical reasoning is that the Japanese employment system makes the introduction of new labor-saving technology very easy. Because of the lifetime employment system, it is said, Japanese workers do not fear being laid off. Some observers point to the Japanese compensation system as another factor. Employees at major firms typically receive a third or more of their pay in the form of bonuses. If the firm is doing badly bonuses may be cut; if it is doing well they may be increased. This is supposed

19. As this was written, Westinghouse Electric bought control of Unimation.

20. Iinuma, "Robotto Kenkyu no Sai Zensen," p. 114.

21. Jerry W. Saveriano, "An Interview with Joseph Engelberger," *Robotics Age,* 3:10-23 (Jan.-Feb. 1981).

to give added incentive to Japanese workers to accept change. Still another factor is the Japanese enterprise union. The argument is that since the future of the union in Japan is bound with that of the company, the union is more likely to accept technological change that it sees as necessary for the long-term survival of the firm.

In large measure this rosy picture may be valid, and we should certainly learn what we can from it, but it should be pointed out that there are a few problems.

First, even though they may be ahead of us, the Japanese have just begun to feel the impact of robotics. In 1980 Japan produced about three times as many robots as the United States, some 3000. But some projections suggest that in 1990 they will produce 60,000 robots. One firm alone reportedly plans to have as many as 100,000 robots in service. In sum, at this point only a very small percentage of the ultimate uses for robots are being exploited.

Second, we sometimes get the impression that the entire Japanese work force is protected by lifetime employment. This is not true. It is not true of most female employees, nor is it true of some two-thirds of the male work force.[22]

The biggest group excluded from the lifetime employment system is women; and there are some indications that women are bearing some of the brunt of the shift to robotics. Women are traditionally expected to leave their jobs when they marry. This allows firms to reduce their work force quickly by attrition; women are not hired to replace those who leave. Many women may never be able to enter the work force, though they are never counted among the unemployed. Moreover some Japanese housewives work part-time to earn extra money at jobs at electrical appliance and other plants; many of these jobs may be disappearing. Again the resulting hardships may not be apparent from employment statistics.

Among the two-thirds of the male work force left out of the lifetime employment system are the employees of smaller firms. We have noted that several Japanese policies have been specifically designed to speed the use of robots by smaller firms. Small and medium-sized firms, for example, reportedly account for some 80 percent of all the welding robots in use. Moreover articles have started to appear in the Japanese press questioning the impact of robotics on smaller firms and their employees. One article describing the situation at a small plastic mold and painting factory said that in October 1980 the company installed two painting robots, which were leased for around $2500 per month. The company was pressured into using robots by a major electric appliance maker that took the bulk of its output, and any savings realized were passed on to this larger company. At the same time the welders and others working at the small firm were engaging in a John Henry-style race with the robots to keep their jobs.[23]

Even the elite of the Japanese work force, male lifetime employees of major

22. Robert Cole, *Work, Mobility and Participation* (Berkeley: University of California Press, 1979), p. 61. This is a very rough estimate. It includes the self-employed among the male work force. If only those employed by others were included, about half the male employees might be considered protected by permanent employment.

23. Keiji Ikehata, "Robotto sai zensen o miru," *Ekonomisuto*, 10 Apr. 1981, pp. 60-65.

firms, are not completely sanguine about their future in an age of robots. Union officials and other observers show increasing concern.

At their 1981 national conventions, unions of electrical and automobile workers called for policies to cope with robots on production lines. They called for greater participation on the part of labor in decision making related to the introduction of new technology. In September 1981 the Japan Council of Metal Workers' Unions adopted an action program to smooth out the impact of robots on their members. So far the concern is not so much unemployment as other consequences of robotics. A survey of some 350 manufacturing companies employing members of the Metal Workers' Unions indicated that middle-aged and older workers had difficulties learning how to operate robots and other advanced equipment, and adjusting to new jobs when their old jobs were taken by robots. The union negotiated an agreement with an electric company whereby workers 45 or older would not be transferred to other locations when robots were introduced.

Unions can do little, however, to help members who are reaching their late fifties, the retirement age at most major firms. For most workers retirement means the loss of a well-paid job, and benefits inadequate to cover living expenses. Typically a less well-paid job is taken at a small firm to make up the difference. The small firms, however, are rushing to replace workers with robots.[24] In 1982 the *Japan Times Weekly* cited unidentified nongovernment statistics projecting that the introduction of robots would cause the Japanese unemployment rate to rise from 2.3-2.5 percent in 1982 to 12 percent in the early 1990s.[25] Others discount the fear of unemployment, but there is sufficient concern that in 1981 and 1982 the Japanese government initiated several technology assessments of the impacts of robots.[26]

LESSONS FROM JAPAN?

The challenge from Japanese robot producers and users is clear: the Japanese are quickly coming to dominate this strategic new industry. This domination seems likely to feed the competitive strength of many other Japanese industries. It is unlikely that we can or should ape all the measures the Japanese have used to foster their robotics industry; some of these would conflict with important American values. It does seem, however, that we should be looking at the Japanese experience for ideas. Over much of their history the Japanese have assimilated and rearranged foreign ideas to make them consistent with Japanese values. Perhaps the United States should have some analogue of the Japanese Large-Scale Projects, coordinated by the government and exempt from antitrust statutes. Perhaps we need tax and other policies to encourage the use of advanced technologies. Maybe we need government involvement in pro-

24. *The Micro-Electronic Revolution and Labour Management Relations* (Tokyo: Japan Productivity Center, 1982). See also Masaru Fujino, "Response of Labor Unions to Technological Progress," *Japan Labor Bulletin,* 1 July 1982, pp. 5-8.

25. "Workers Begin to Re-appraise Robotics Revolution," *Japan Times Weekly,* 17 Apr. 1982.

26. Studies were undertaken, for example, by the Employment Policy Study Group of the Ministry of Labor, the Long-Range Prospects Committee of the Economic Advisory Council of the Prime Minister, and MITI.

grams to enable small businesses to use new technology more quickly.

While many have overstated the success of the Japanese in ameliorating the social costs of introducing such new technologies as industrial robots, it would be a mistake to ignore the possibility of learning from a country that has had over a century of success in coping with the problems associated with rapid technological change. The Japanese will have many advantages in dealing with the social impacts of robotics.

One advantage they will probably have during the transition to robotics is their system for industrial education. The education system is closely tied to business and industrial policy in Japan. In the late 1950s and early 1960s, for example, when Japan undertook its successful gross national product doubling program, education was a primary concern. Business organizations and government advisory groups made up of businessmen, scholars, and government officials studied the vocational education systems of other countries and looked at the projected human resource needs for the newly enlarged Japanese economy. They concluded that there would soon be a need for some 440,000 additional technicians—people trained at the high school or junior college level. Programs were initiated, often jointly by schools and businesses, to ensure the supply of these technicians. New technical junior colleges were set up, some of them by such large business firms as Hitachi. In this and other rapid changes in the employment structure since the Meiji Restoration in 1868, the Japanese appear to have been highly successful.[27]

More generally Japanese industrial workers tend to have a better education in mathematics and science than their American counterparts. This will presumably enhance their retrainability.

There has been far less serious research on company training systems in Japan than would seem justified by the importance of the Japanese experience, but it appears that Japanese firms are much more thorough in training and retraining their employees than their U.S. counterparts. Obsolescent skills are replaced—presumably with some pain, as middle-aged and older workers give up comfortable niches—but at least without the problem of unemployment. Robert Cole, for example, describes workshop universities at one of the Toyota companies, and Ronald Dore describes the intensive in-service training at Hitachi. These systems may be possible in Japan because the lifetime employment system allows firms to recover much of the investment they make in developing their employees. The United States may have to find alternative ways to continue the development of its people.[28]

When interfirm transfers become necessary in Japan, the prefectural and national governments attempt to retrain people. In some instances when industries have been contracting, as a few years ago in shipbuilding, employers transfer excess personnel to firms in other industries, subsidizing their pay if

27. Nobuo Shimabara, *Adaptation and Education in Japan* (New York: Praeger Publishers, 1979).

28. Cole, *Work, Mobility and Participation*, pp. 180-86, and Dore, *British Factory-Japanese Factory* (Berkeley: University of California Press, 1973), pp. 62-67.

necessary, so that employees entering a lower-paying job do not suffer an actual reduction in pay.

In these areas we should be examining the Japanese experience and seeing what elements of Japanese educational and training policy should be adapted for use in the United States. Does the United States need ongoing business-government councils on changing educational needs? Should we attempt to develop Japanese-style intersectoral job-transfer systems?

The caveat is that the tremendous pressure on U.S. government and business to meet the Japanese challenge may be dangerous. One danger is a rejection of the challenge—for example, the use of political solutions such as trade barriers to avoid the more difficult process of learning how to become more competitive. Another is the widespread borrowing of practices that are not well understood. We know little about just how Japanese policies work, why they work, and what consequences they entail. In this as in so many other areas, we as a nation need very quickly to improve our ability to learn from Japan, and the rest of the world.

ANNALS, *AAPSS*, **470**, November 1983

Robotic Realities: Near-Term Prospects and Problems

By ROBERT U. AYRES and STEVEN M. MILLER

ABSTRACT: Industrial robots are automation, but with a difference. Other machine tools are extensions of human capabilities, while robots are seen mainly as substitutes for human workers. Robots will find most of their industrial applications during the next decade or two in the metal-working sectors, where they will begin to displace semiskilled machine operatives in medium to large batch production operations. They cannot substitute for skilled machinists or other workers doing nonroutine jobs, or specialized, dedicated hard automation used in mass production. The current generation of robots, lacking sensory data processing and interpretation capabilities, can potentially replace up to 1.3 million manufacturing jobs. The next generation, with crude vision or tactile senses; will potentially displace about 3 million more. However, only relatively large firms can profitably utilize many robots at present; it may be 20 years or more before these usage rates are achieved in practice. A shift from stand-alone machine tools, to manufacturing cells consisting of several machine tools served by a robot and controlled by a computer, will accelerate the practical use of robots in the 1990s.

Robert U. Ayres, professor of engineering and public policy, Carnegie-Mellon, has research interests in economics, technology, public policy, and methodological issues. He has written or coauthored six books, including Technological Forecasting and Long-Range Planning *(1969);* Uncertain Futures *(1979); and with Steven M. Miller,* Robotics: Applications and Social Implications *(1983).*

Steven M. Miller is assistant professor of engineering and public policy and industrial administration, Carnegie-Mellon, and a member of the university's Robotics Institute. His areas of interest are the economics of technology and the impacts of new technologies on the organization of production. For the past two years he and Dr. Ayres have collaborated on research into the economic and social implications of flexible automation, especially robotics.

NOTE: This article is based largely on material that has been previously published. Major sources include Robert U. Ayres and Steven M. Miller, *Robotics: Applications and Social Implications* (Cambridge, MA: Ballinger, 1983); idem., "Robotics and Conservation of Human Resources," *Technology and Society,* 4(3):181-97 (Winter 1982); idem., "Socio-Economic Impacts of Industrial Robotics: An Overview," in *Handbook of Industrial Robotics,* ed. Shimon Y. Nof (New York: John Wiley & Sons, Inc., 1984, forthcoming). Reprinted by permission.

INDUSTRIAL robots are machine tools. In December 1982 there were about 33,000 in use worldwide, nearly 60 percent of them in Japan. About 7000 robots were in U.S. factories. Robots are not human-like androids that can walk around and converse, as in literature and films. Most of them are permanently fixed in one location. More realistically they are programmable manipulators that can move parts or tools through a prespecified sequence of motions. They are also—at present—crude, clumsy, blind for the most part, and very stupid. They usually have one arm and several steel fingers with a total of four, five, six, or at most seven degrees of freedom. Reprogrammability means only that the robot's actions can be modified by changing control settings, without changing the hardware. Robots combine some attributes of traditional machine tools as well as of machine operators. Like a machine tool they can repeat the same task for prolonged periods with great precision. By definition a robot must be flexible enough to be taught to do a new task, and to use accessory tools to extend its range of physical capabilities. The following is the classification of robots used in Japan; the first two categories in this Japan Industrial Robot Association (JIRA) classification are not currently defined as robots in the United States.[1]

1. Manual manipulator. A manipulator that is directly operated by a man.

2. Fixed sequence—pick and place—robot. A manipulator that performs a sequence of steps specified by means of cams, mechanically set limit switches, or hydraulic or pneumatic valves.

3. Variable sequence robot. A manipulator as defined in item 2, for which instructions can be specified by resetting electrical connections.

4. Playback robot. A manipulator that can remember and repeat any operation after being instructed by an operator; subcategories are known as walkthrough and leadthrough.

5. Numerically controlled (NC) robot. A manipulator that can receive instructions in digital form via magnetic tape or directly from a computer.

6. Intelligent robot. A robot that can modify its own actions in response to data obtained through its sensing and recognitive abilities.

Robots are utilized in industry for the usual virtues of machines: strength, speed, precision, untiring availability, predictability, reliability, and relative imperviousness to hostile environments. They do not as yet possess several important natural human capabilities: the ability to react to unforeseen circumstances or changing environments, and the ability to improve performance based on prior experience.

Thus, while robots are classed as flexible automation, this characterization is relative. Robots at present are flexible only in the limited sense that a programmed sequence of actions can be altered without actually rebuilding the device. For the crudest devices—types 1, 2, and 3—the reprogramming is actually done manually, by resetting switches or valves. For playback robots, type 4, the reprogramming is normally done by physically moving the robot arm through the desired sequence of motions. In the case of NC robots, type 5, instructions are digitized and read electronically from a magnetic tape or via some communications interface, from an external con-

1. JIRA, *The Robotics Industry in Japan: Today and Tomorrow* (Englewood Cliffs, NJ: Prentice-Hall, 1982).

FIGURE 1
THE RELATIVE FLEXIBILITY AND EFFICIENCY OF ROBOTS, HUMANS, AND SPECIALIZED MACHINES

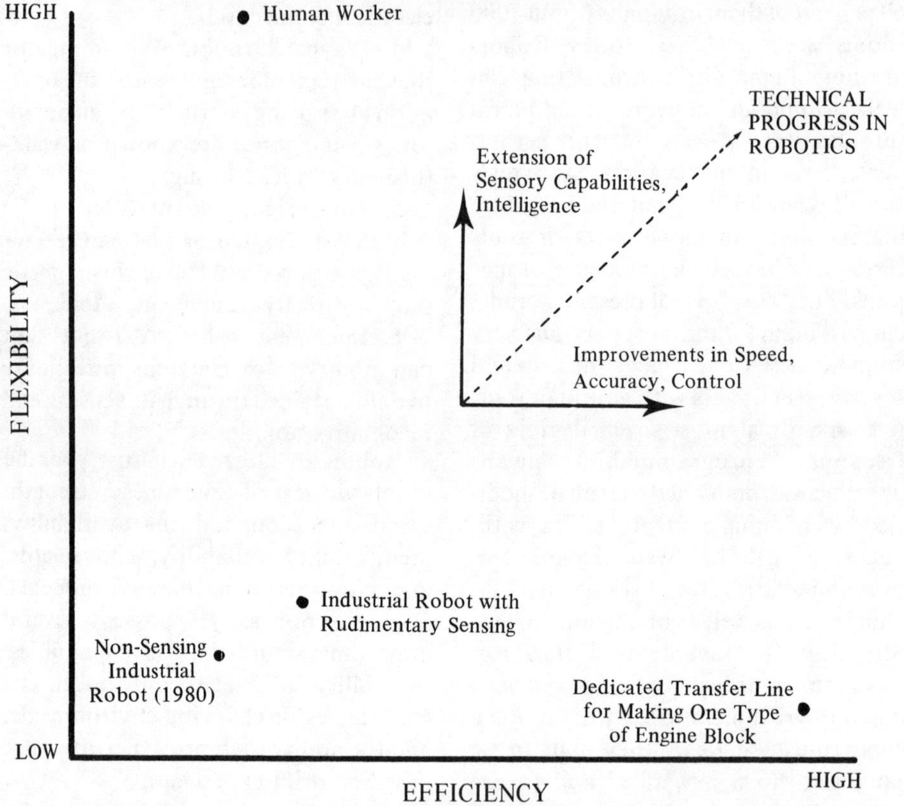

troller-computer. State-of-the-art robots, type 6—mostly in research labs—do have crude senses of sight and touch, and limited capability to coordinate their manipulators with sensory input. Instructions must be provided externally, however, by a programmer.

Robots are flexible only when compared to so-called hard automation. They are not as yet particularly good at adjusting to varying conditions and unpredictable circumstances, as human workers can. Roughly the flexibility of humans is derived from a combination of high order sensory capability—vision, touch—and problem-solving ability—intelligence. On the other hand the purely mechanical attributes—strength, speed, precision—that robots share with other machines can be termed dexterity. Not surprisingly, while humans far exceed robots in sensitivity and intelligence, they lag somewhat in dexterity, as Figure 1 suggests qualitatively. Apparently biological organisms use intelligence and sensitivity to compensate for the inherent limitations of organic materials. Essentially the tasks robots can do better than humans are those requiring more dexterity than flexibility.

Because of current limitations, today's robots are usefully employed in highly structured industrial environments where practically all of the variability and decision making can be engineered out of the workplace. Existing uses of industrial robots all involve repetitive preprogrammable tasks such as spot welding, spray painting, palletizing, and the loading and unloading of metal forming, metal cutting, and other types of machines. The next generation of sensory-based robots will be able to perform a broader range of tasks under less structured conditions, in addition to becoming cheaper and easier to use. Expected uses of robots with vision and improved feedback control will include inspection, assembly, heat treatment, grinding and buffing, and electroplating. Capabilities of commercially available robots, and capabilities under development for future robots are listed in Table 1.

Eventually many of the hands-on tasks performed by production workers on the factory floor will be done by robots in computer controlled manufacturing systems. Programmable automation is beginning to replace the current generation of manually controlled machines. This transition will undoubtedly continue for many decades. There is potential for significantly improving the productivity of the manufacturing sector and increasing the wealth-producing potential of the economy as a whole. We also face significant social impacts, such as the short-term prospects of technological displacement, and the longer-term prospects of basic structural shifts in the economy.

APPLICATIONS OF INDUSTRIAL ROBOTS

Specific tasks for which robots are used today are shown in Table 2. Experimental and future applications are also suggested. A summary characterization of benefits is shown in Table 3. Robots are not really cost-effective in most custom manufacturing applications, because in such cases a large fraction of the labor time is spent setting up the machines. Since each job is unique the setup requires the active involvement of a skilled machinist. For one-of-a-kind—prototype— or few-of-a-kind products, it is easier for a skilled machinist simply to make the piece than to figure out how to teach a robot or program an NC machine to do it. Thus custom production generally requires skilled workers and relatively inexpensive, manually operated, general purpose, single-spindle machine tools. All operations are carried out sequentially.

Robots are not generally cost-effective in mass production applications either, because specialized mass production machinery can usually perform the operations more efficiently than general purpose NC machines. Mass production machinery, or hard automation, is ultraspecialized to repeat a fixed sequence of operations at high speed for very long periods of time. Many operations are carried out in parallel on specially designed and dedicated transfer machines, perhaps with 100 or more tools—for example, drills—cutting simultaneously. Auto engines and transmissions are manufactured in this way. It is always difficult and expensive if not impossible to reconfigure the hard automated system for another product. It is usually cheaper to scrap the specialized

TABLE 1
ROBOT CAPABILITIES

	Commercially Available Capabilities (1980)	Capabilities Sought for the Future
Learning	on-line programming via teach/playback modes	general purpose robot programming languages
	teaching in multiple coordinates	off-line programming
	local and library memories of any size	learning with experience
Decision making	program selection by random stimuli	world model of working environmental
Sensing	computer interpretation of sensory data	positional sensing
	computer interfacing	three-dimensional vision with gray levels and color
	two-dimensional vision with binary recognition	tactile sensing
	force/torque sensing	voice communication
	limited speech input	improved processing of sensory inputs
		coordination of multiple sensory inputs and control
Manipulation	six infinitely controllable articulations between base and gripper	miniature manipulators
	point to point control	greater position accuracy
	continuous path control	greater dynamic control
	position accuracy repeatable to 0.3 millimeters	general purpose hands
	handles up to 150 kilos	multiple hand-to-hand coordination
Mobility	synchronization with moving workpieces	programmable omni-directional mobile bases
		self-navigating mobile bases
		walking robots
Reliability	400 hours for mean time between failures	self-diagnostic fault tracing

SOURCE: Adapted from Joseph F. Engelberger, *Robotics in Practice* (New York: American Management Association, 1980); see also Robert U. Ayres and Steven M. Miller, "Industrial Robots on the Line," *Technology Review*, 85(4):35-46 (May-June 1982).

machinery and to build a new system from scratch.

Between the two extremes—custom and mass—lies the domain of batch production, where NC machine tools and robots are particularly appropriate. It may be noted that very complex assemblies with many parts tend to be

TABLE 2
ROBOT APPLICATION AREAS

Commercially Available Robots	Experimental Robots	Future Robots
Die casting	Assembly:	Exploration and extraction of materials:
Spot welding	electromechanical	in subsurface mines
Arc welding	insertion of other small fragile components	underseas
Investment casting	Inspection:	Satellite repair
Forging	of electromechanical parts and assemblies	Space manufacturing
Press work	of electronics parts and assemblies	
Spray painting	Sheep shearing	
Glass manufacturing		
Plastic molding		
Foundry work		
Machine tool loading		
Heat treatment		
Deburring of metal parts		
Palletizing		
Brick making		
Dimensional inspection		

produced in smaller batches than simpler items.[2] The typical relationship between the number of different parts made in a factory and output rate or batch size for each part is shown schematically in Figure 2.

For custom and small batch production, the cost of a product is predominantly attributable to labor, and unit costs do not decline much with volume. As batch size increases, more advanced—for example, numerically controlled—machines can be justified, since programming costs and specialized tools and jigs can be spread over more units. But the labor content is lower and unit costs decline more rapidly with volume.

At still larger production volumes it is possible to justify still more labor-saving specialization, including the integration of one or more robots and several machine tools in a flexible—reprogrammable—computerized manufacturing cell as depicted in Figure 3. Capital costs are higher, but unit costs drop still faster. With true mass production the flexibility or reprogrammability is sacrificed in exchange for maximum output rates. Typical total cost versus quantity schedules are shown in Figure 4, and unit cost versus quantity is shown in Figure 5. Note that Figure 5 shows three overlapping regions, characterized by different basic production technologies, discussed previously.

Each of the three technologies is the cost-minimizing choice only within the volume range for which it is intended, and an inefficient choice outside its appropriate range. For a given product, unit cost would decrease regularly as output increases over a wide range of volumes if one considers the lower envelope of the long-run cost curve, where the

2. The automobile appears to be an exception to the rule but actually is not. The final product, however, is assembled in moderately small batches from simpler subassemblies—engine, transmission, wheels, full system, brakes, and so on—that are themselves mass produced.

TABLE 3
CHARACTERIZATION OF ROBOT BENEFITS

	Current Industrial Robots	Experimental Industrial Robots	Future Robots
Areas of application	Highly structured manufacturing task	Variable manufacturing task	Subsurface Undersea Space
Description	General purpose machine tool with no sensing capabilities	General purpose machine tool with limited sensing and decision-making capabilities	Autonomous system guided by artificial intelligence
Benefit	Factor saving: capital material labor	Factor saving: capital material	Resource extending

optimal—cost minimizing—technology is used for each level of output.

About 90 percent of current robot users in the United States fall within the industries of the metalworking sector. As the name implies, these industries are engaged in the fabrication, finishing, or assembly of products from standard metal shapes, and from mechanical and electronic parts and subassemblies purchased mostly from other metalworking industries. The metalworking sector includes the following groups of industries, designated by standard industrial classification (SIC) codes.[3]

—34: fabricated metal products;

—35: machinery, except electrical machinery;

—36: electrical equipment and machinery; and

—37: transportation equipment.

These industries have been described by Thomas Vietorisz as "the bellwether of economic growth"[4] for an industrial society because they produce all of the tools and capital equipment used by all manufacturing industries and all other sectors of the economy. The metalworking sector is the locus within the industrial system where new technological knowledge is embodied in a physical form, especially as capital goods. Since capital goods play such a critical role in the creation of new products, processes, and wealth, the importance of this sector extends far beyond the number of people it directly employs.

Based on an analysis of unit processing cost and level of output, we have estimated the dominant mode of pro-

3. This system of industrial classification has been developed over a period of many years under the guidance of the U.S. Department of Commerce. All data collected by the Bureau of the Census as part of the economic census, including the Census of Manufacturers, is organized according to the SIC system.

4. *UNIDO Monographs on Industrial Development: Volume 4, Engineering Industry* (Vienna: U.N. Industrial Development Organization, 1969).

FIGURE 2

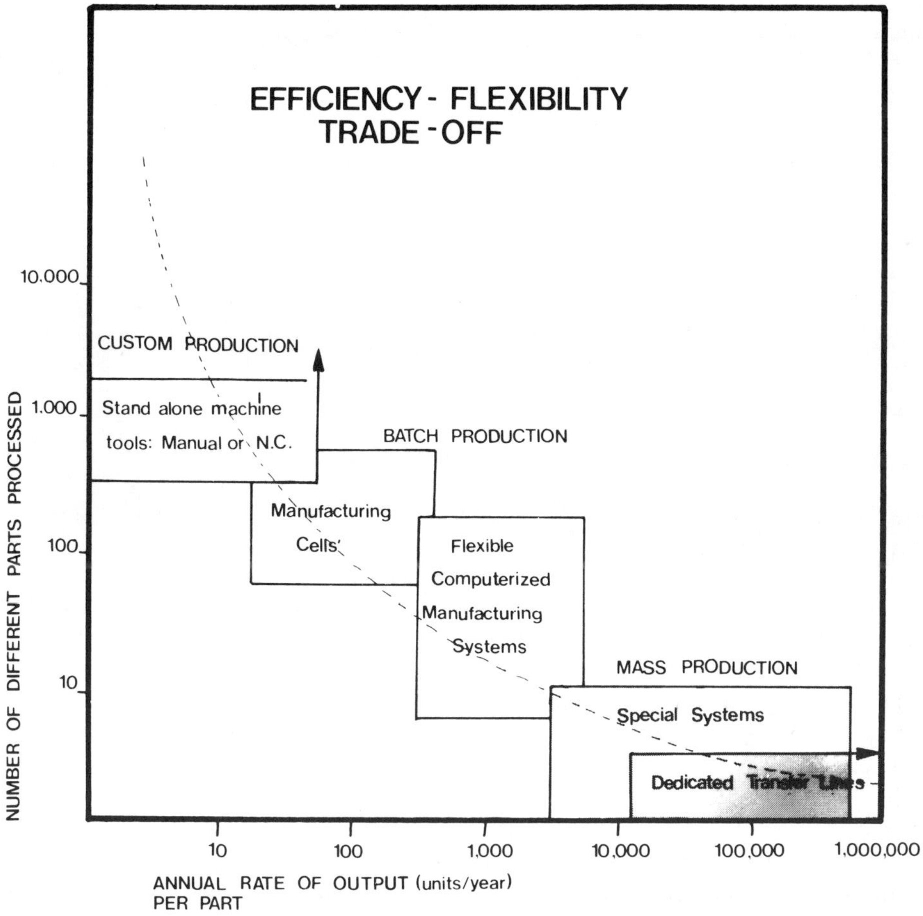

SOURCE: Reprinted with permission from Ayres and Miller's *Robotics: Applications and Social Implications.* Copyright © 1983, Ballinger Publishing Company.

duction within 101 metalworking industries in SICs 34-37. Within this group the highest unit processing cost per unit output, measured in mass, is for SIC 3662, radio and television communication equipment, an extreme case of custom or very small batch production. The lowest unit processing cost is for SIC 3465, auto stampings, an extreme case of mass production. The range between these two extremes is a factor of 60. That is, SIC 3662 spends over 60 times as much as SIC 3465 in capital and labor cost per pound of material processed. Some of the unit cost differential between these two industries is due to differences in product complexity. However, if the unit processing cost of auto stampings is compared to that of an equally complex type of product that is custom produced such as nonferrous forgings, there is still a tenfold difference.

FIGURE 3
COMPUTER-ROBOT INTEGRATED MANUFACTURING CELL

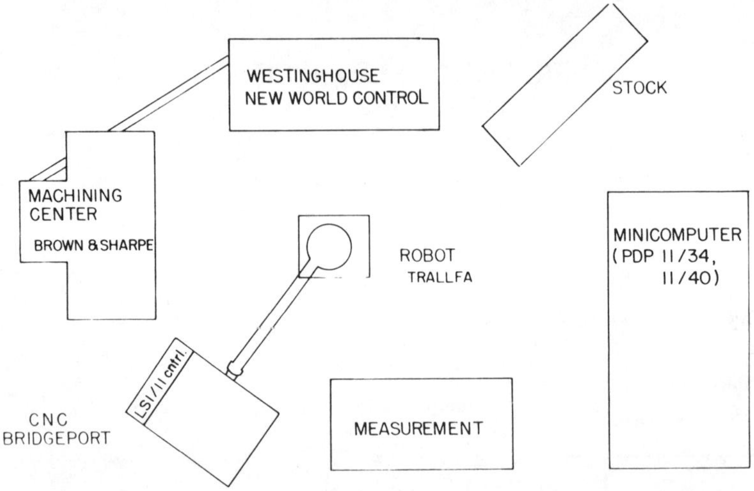

As a rough generalization, for each 1 percent increase in the level of output, unit processing cost decreases by 0.38 percent for the metalworking industries in our sample. This implies that if a typical product were produced one-of-a-kind, unit processing cost would be 200 times greater than if it were mass produced at a million copies per year. If a typical product were produced in batches of 100 or 1000, unit processing cost will still be 35 to 15 times higher than if it were mass produced at a million copies per year. Clearly producers and consumers are paying a high premium for metal products produced in small batches.

We estimate that mass production industries account for about one-quarter of the value added, and about one-third of the total output in the metalworking sector. This suggests that much of the value added can be thought of as the cost of flexibility needed for product differentiation and specialization.

There are a limited number of robot applications in manufacturing sectors other than metalworking, though at present the problems associated with processing nonrigid or delicate materials, and with very high-speed production lines, restrict their use. Current and near-term future applications of robotics in the processing of leather, rubber, asbestos, plastics, and food, and in the manufacturing of glass, clothing, and wood products are briefly reviewed in R. D. Schraft, E. Schults, and P. Nicolaisen.[5] In both Japan and the United States demand for robots in the nonmanufacturing sectors of the economy currently account for a negligible part of the total market.

POTENTIAL IMPACTS ON LABOR COST

Existing and likely near-term capabilities of robots make them candidates

5. "Possibilities and Limits for the Application of Industrial Robots in New Fields," in *Tenth International Symposium on Industrial Robots, Milan* (Bedford, England: IFS Publications, 1980).

FIGURE 4
TOTAL COST VERSUS QUALITY SCHEDULES FOR THE MANUFACTURE OF A PARTICULAR PART BY DIFFERENT PROCESSES

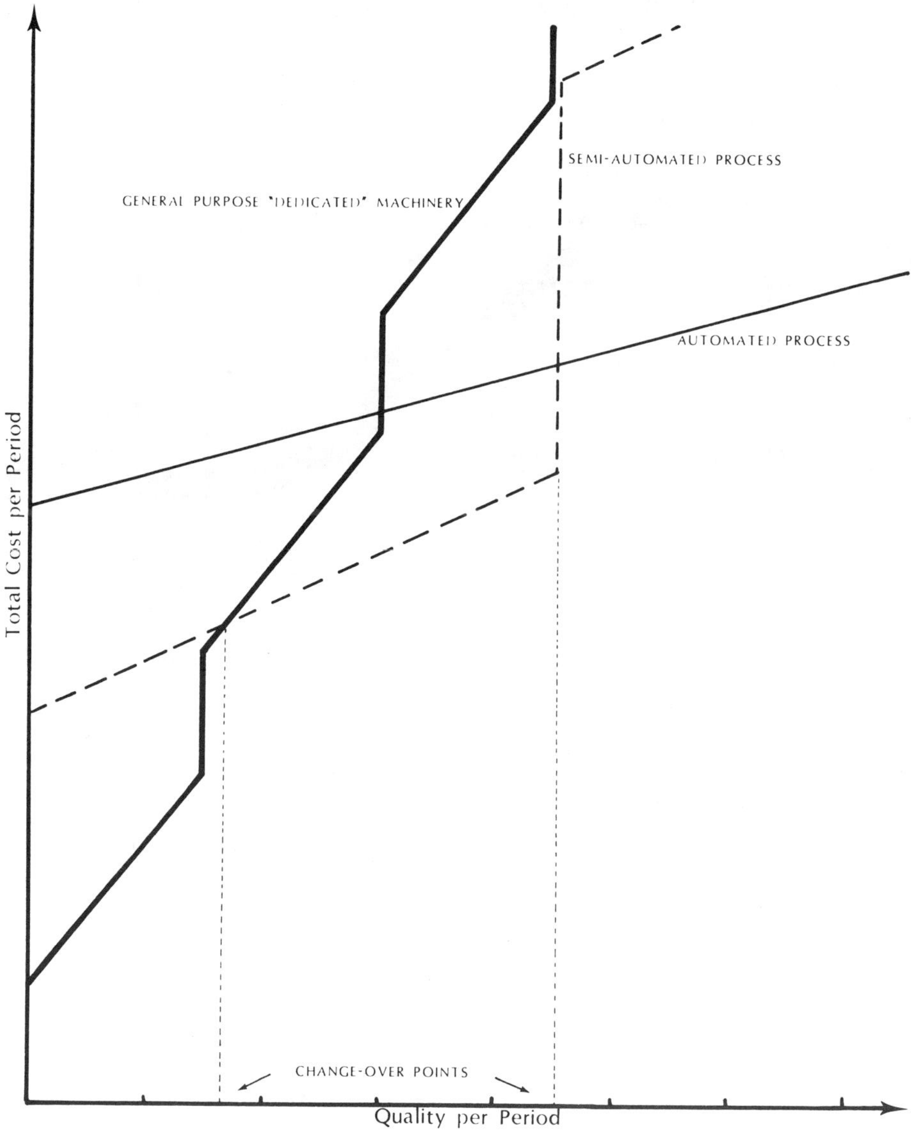

to replace significant numbers of machine operators and unskilled laborers over the next three decades or so. In general terms the replacement of several million production workers by robots will decrease direct labor input. Consequently output per labor hour will most certainly increase, although the effects on total factor productivity depend on the requirements for new capital and other ways in which robotic—and computer aided design/computer aided manufacturing (CAD/CAM)—technology may alter production technology.

Within the metalworking industries, where robots seem to be most applica-

FIGURE 5
UNIT COST VERSUS QUANTITY

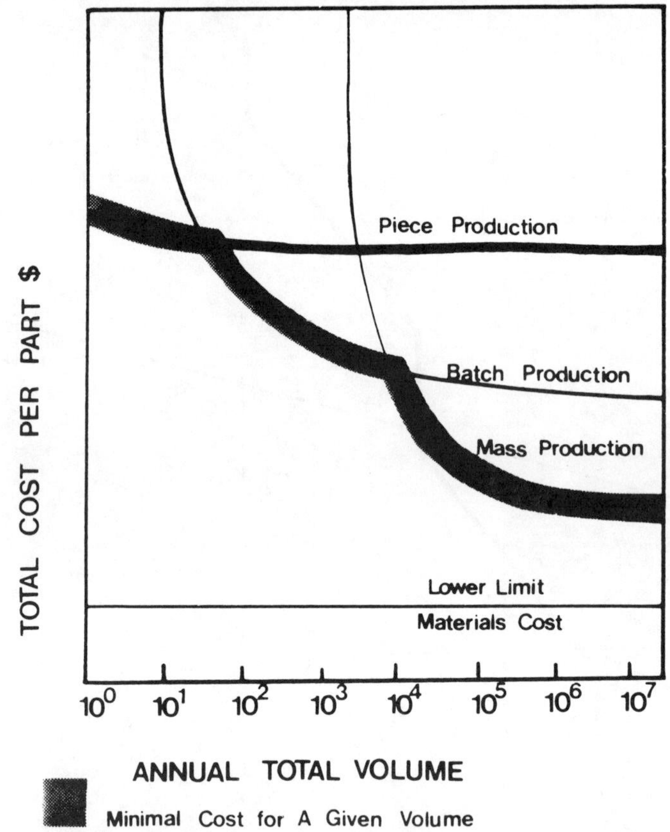

SOURCE: Reprinted with permission from Ayres and Miller's *Robotics: Applications and Social Implications.* Copyright © 1983, Ballinger Publishing Company.

ble, direct wage and benefit payments to production workers comprise less than 18 percent of the total value of output on average. The true labor cost is actually the sum of direct labor cost plus the labor cost embodied in the purchased materials. However, only direct labor costs are relevant to technology choice. Hence for now we consider an industry in isolation, without considering the effects of cost savings that might be passed on to customers.

Potential reductions in production labor is based on three scenarios, outlined in Table 4. Fabrication workers include all types of skilled and semiskilled machine operators and setup workers, as well as material handlers, laborers, and miscellaneous types of skilled, semiskilled, and unskilled production workers. It is assumed that maintenance workers will not be replaced. The low scenario represents the current potential for replacing factory workers with insensate robots. The medium scenario represents the near-term potential for replacing factory workers with the emerging generation of

TABLE 4
REPLACEMENT SCENARIOS

	Percentage Replaced		
Occupation	Low	Medium	High
Fabrication workers	20	50	75
Assembly workers	0	25	75
Inspectors	0	0	75
Supervisors	0	0	75

sensor-based robots. From a technical standpoint these two scenarios could be realized within the decade. The high scenario is our own subjective estimate of the long-term potential for eliminating production labor in the millenial factory of the future.

In the low- and medium-replacement scenarios, the potential direct cost reductions—not counting pass-through effects from suppliers—appear to be modest, averaging near 2 percent and 5 percent, respectively. The potential cost reduction is greater in the high-replacement scenario, averaging about 14 percent. Of course, the upper limit on potential cost savings from eliminating production workers is given by the total production worker portion of output in the near term, considering the low- and medium-replacement scenarios. It does not appear that the direct substitution of robots for factory workers by itself would have a substantial impact on total cost even in those industries where robot use is most heavily concentrated. Something else must happen if large savings in cost and increases in productivity are to be realized in the near term.

POTENTIAL IMPACTS ON CAPACITY

Possibly more significant is the impact of robotization and CAM on factory organization and utilization of machinery and equipment. Throughout the major metalworking industries there are on average more machines than operators. The actual ratio of machines to operators understates this point, since each machine is available three shifts per day, whereas an operator only works a single shift, or slightly longer, allowing for overtime. Comparing the total hours worked by machine operators to the total hours machines are available, it is clear that on average machines are operated only for a small portion of the total time they are available. According to our estimates in Table 5 the effective utilization of manually operated metal cutting and metal forming machines and welding equipment is remarkably low.

The average figures are 13 percent, 14 percent, and 21 percent, respectively, assuming that theoretical utilization corresponds to 24 hours per day, seven days a week, with one operator per machine. On average, NC metal cutting machines are utilized more fully than manually controlled cutting and forming machines. However, in 1977 fewer than 3 percent of the metal cutting machines were NC. These estimates measure only the proportion of time an operator is available to run the machine, with no allowance for the possibility that the machine may be idle part of the time when the operator is on duty. Produc-

TABLE 5
ESTIMATES OF AVERAGE MACHINE TOOL UTILIZATION IN THE METALWORKING INDUSTRIES, 1977*

Major Group (SIC)	Machine Type (percentages)				
	Metal Cutting, Manual	Metal Cutting, Numerically Controlled	Metal Forming	Joining—Welding	All Machines, Average
34	12.4	18.4	15.8	15.1	14
35	12.7	20.9	7.4	17.8	13
36	9.9	29.3	13.0	8.2	11
37	16.5	14.8	17.5	40.0	20
Average	12.9	20.0	13.6	21.3	14

SOURCES: Machine tool hours derived from "The Twelfth American Machinist Inventory of Metalworking Equipment, 1976-78," *American Machinist*, 122(12):133-48 (Dec. 1978); labor hours derived from Bureau of Labor Statistics, *Occupational Employment in Manufacturing Industries, 1977* (Washington, DC: Government Printing Office, 1980).

*Machine utilization is defined here as follows:

$$\text{utilization} = \frac{\text{total operator hours available}}{\text{total machine hours available}}$$

total operator hours available = (number of operators) × (average hours worked per operator per year)

$$\text{average hours worked per operator per year} = \frac{\text{total hours worked by production workers}}{\text{total production workers}}$$

total machine hours available = (number of machines) × (8760 hours per year)

8760 hours per year = (24 hours per day) × (365 days per year)

tive cutting time by machine tools as a fraction of theoretical capacity in low- and medium-volume shops is 6 percent and 8 percent, respectively, increasing to 22 percent for high-volume, mass production operations.

Thus it is clear that most machines, especially in the metal working industries, are idle most of the time, even before making allowances for setup time, load/unload time, and other adjustments. Even in the most automated industries, such as the production of automobile engines and transmissions, true machine tool utilization rates as high as 50 percent are seldom if ever achieved. We believe the major quantifiable economic impact of robotics and CAM will be to expand sharply the effective capacity of production facilities by increasing both the amount of time per year the plant is operating and the throughput per shift. Much of the lost time is due to incomplete use of the second and third shifts and to plant closings for holidays, strikes, and other reasons. Weekends, holidays, and night shifts are less popular than the normal 40-hour work week. Consequently even if labor is available during these periods, it is more expensive.

Under normal operating conditions in a healthy economy even high-volume plants are shut down nearly 80 days per year due to Sundays, holidays, and planned closings for retooling. Mid-volume plants are closed on average 102 days—all weekends—and low-volume plants are closed nearly 125 days out of the year—weekends plus three weeks for

TABLE 6
POTENTIAL PERCENTAGE INCREASES IN OUTPUT FROM UTILIZING LOST TIME

Type of Plant	From Utilizing Days Plant Is Closed	From Utilizing Nonscheduled Production Time	Total Increase in Output
High-volume	28	3	31
Mid-volume	83	115	198
Low-volume, one-shift operation	148	187	335
Low-volume, two-shift operation	74	43	117

SOURCE: Corrected from Robert U. Ayres and Steven M. Miller, *Robotics: Applications and Social Implications* (Cambridge, MA: Ballinger, 1983); see also Steven M. Miller, "Potential Impacts of Robotics on Manufacturing Cost within Metalworking Industries" (Ph.D. diss., Carnegie-Mellon University, 1983).

holidays and shutdown. When open for production, high-volume plants are typically operating over 22 hours per day, whereas mid-volume and low-volume plants are typically scheduled to operate 10.7 and 8 hours per day respectively. Clearly there is considerable potential for increasing output, and thereby decreasing unit cost, by saving long runs for an unmanned third shift or weekend, using robot operators. During the next two decades, as manufacturers gain experience with unmanned factory operations, the less routine machine setup, repair, maintenance, and inspection tasks could be reserved for the regular day shift. Planned shutdowns required for retooling would be substantially reduced, or possibly eliminated in a robot-integrated factory with flexible production technologies.

A summary of the potential for increasing available production time, hence output, in existing facilities is shown in Table 6. Output per year could be increased by perhaps 30 percent in high-volume plants, and by at most 200 percent in mid-volume plants. If low-volume plants are operating only one shift per day, recouping cost time could increase output by over 330 percent. If we generously assume that low-volume shops are operating on a two-shift basis—which only a portion do—output per year could still be increased by nearly 120 percent.

In addition to extending the amount of working time per year, robotics, especially when integrated with other CAM technologies, can increase capacity by increasing the number of parts produced per hour. As we have seen in Table 5, machines are productively engaged only about 30 percent of the time the plant is open and operating. The remaining 70 percent of possible production time is lost for a variety of reasons. Recouping the fraction of possible production time that is lost to nonproductive uses would further increase the effective capacity of machine tools.

Much of the lost time is machine related: equipment limitations, tool changing, and equipment failures. However, a sizable fraction of time lost is due to management and work force practices, including personal time breaks, late starts, early quits, material handling, excessive machine adjustments, and in-line storage losses due to scheduling inefficiencies. Personal time, late starts, early quits, and some fraction of the material handling time could be virtually eliminated by replacing workers with robots. Time losses due to tool changing, equipment failures, excessive machine adjustments, setups, and scheduling inefficiencies will probably not be affected directly by robots, but might be reduced if more aspects of factory work were consolidated and controlled by sensor-based computer systems. For example, sensors monitoring machine performance would eliminate unnecessary adjustments and speed up diagnosis of machine failures.[6] If stand-alone machines were replaced by a flexible manufacturing system, and if parts processing were rationalized by adopting group technology, there would be less material handling and the scheduling of parts and tools would be simplified. Even a substantial fraction of the equipment-related losses could be eliminated in a fully integrated flexible manufacturing system, since the whole system need not be stopped if one station malfunctions. Robots or programmable pallets under the control of a central scheduling computer could reroute parts to other work stations.

It is difficult to discuss the potential improvements in productivity that may be brought about by robotics in isolation from the development of CAM systems and other forms of factory automation. Retrofitting robots into existing production lines will bring about some improvements, such as improving the utilization of a single machine or work station, but we do not expect that it would dramatically improve overall factory performance. Substantial effects on factory performancc and costs require the integration of robots and other forms of factory automation into coordinated manufacturing systems. It also becomes more difficult and less meaningful to distinguish between robots and other forms of factory automation as the concept of robotics evolves from programmable manipulators to machines and systems that can "sense, think and act."[7]

Our rough estimates of potential increases in throughput that could be achieved from recouping the nonproductive time lost during scheduled operations are based on informed judgments, but have been reviewed by several industry experts. They are not the result of detailed analysis. We distinguish two levels of improvements as a result of (1) uses of robots per se, and (2) integrating robots with CAM systems and other forms of factory automation. We suggest that the installation of robots, without increasing the time normally planned for operations and without extensively adding other forms of automation, would result in a 10 percent increase in

6. In the next few years time lost to equipment failure could conceivably increase as systems become more automated and more complex. However, over the next two decades we expect improvements in machine reliability and sensor-based diagnostic systems to improve machine and system reliability and to reduce equipment failures.

7. We borrow the broader definition of robotics as machines that can "sense, think and act" from Raj Reddy, director of the Carnegie-Mellon Robotics Institute.

TABLE 7
POTENTIAL PERCENTAGE OUTPUT INCREASES FROM WORKING MORE EFFICIENTLY DURING PLANNED OPERATIONS: MID-VOLUME MANUFACTURING

Function	Percentage of Operating Time	Potential Percentage Reduction	Adjusted Percentage	Potential Percentage Reduction	Adjusted Percentage
		(robots only)		(robots with CAM)	
Setup and gauging	22	−30	15.4	−65	7.7
Load/unload and noncutting	12	−40	7.2	−60	4.2
Tool change	22	−5	20.9	−15	18.7
Equipment failure	7	0	7	0	7
Idle time	12	0	12	−25	9
Productive fraction	25	0	25	0	25
Total scheduled production time	100		87.5		64.6
Output index	1.0		1.14		1.55

SOURCE: Breakdown of operating time from John E. Mayer and David Lee, "Estimated Requirements for Machine Tools during the 1980-1990 Period," in *Machine Tool Systems, Management and Utilization,* ed. Arthur R. Thompson (Lawrence, CA: Lawrence Livermore National Laboratory, 1980), vol. 2 of *Machine Task Force Report on the Technology of Machine Tools,* pp. 31-41; estimates of potential reductions from Ayres and Miller, *Robotics: Applications.*

output in high-volume machining operations—not including assembly—and nearly a 15 percent increase in output in mid-volume and low-volume production. If robots were used in conjunction with other forms of factory automation, still without increasing the number of hours normally planned for operations, output might be increased by nearly 50 percent in mid- and low-volume production, and possibly by 40 percent in high-volume production. A breakdown of the mid-volume case is shown in Table 7.

Our judgments regarding the estimated improvements, as in Table 7, are mostly applicable to situations of retrofitting or incrementally adding technologies within existing plants. Other studies suggest that completely redesigning the factory around new technology could result in substantially greater improvements in throughput during planned production periods. For example, John E. Mayer and David Lee[8] of Ford Motor Company estimated the combined effects of applying the most advanced concepts to almost all aspects of machine design and control and factory layout in high-volume machining systems. They considered the use of automatic loading/unloading, automatic tool changing, diagnostic sensing, component reliability, and unmanned operation, as well as improved line balancing and faster cutting speeds. Their results suggest that all of these improvements, without increasing cutting speeds, could result in a 90 percent increase in output. If cutting speeds were increased to their upper lim-

8. "Estimated Requirements for Machine Tools during the 1980-1990 Period," in *Machine Tool Systems, Management and Utilization,* ed. Arthur R. Thompson (Lawrence, CA: Lawrence Livermore National Laboratory, 1980), vol. 2 of *Machine Tool Task Force Report on the Technology of Machine Tools,* pp. 31-41.

TABLE 8
SUMMARY OF POTENTIAL INCREASES IN CAPACITY

Type of Plant	Base Case	Potential Capacity Increases	
		Robots Only	Robots with CAM
High-volume			
Available hour index	1.00	1.31	1.31
Throughput index	1.00	1.11	1.39
Output index	1.00	1.45	1.82
Increase in output (percentage)		45	82
Decrease in unit cost (percentage)*		–13	–21
Mid-volume			
Available hour index	1.00	2.98	2.98
Throughput index	1.00	1.14	1.55
Output index	1.00	3.40	4.62
Increase in output (percentage)		240	362
Decrease in unit cost (percentage)*		–37	–44
Low-volume, single shift			
Available hour index	1.00	4.35	4.35
Throughput index	1.00	1.16	1.52
Output index	1.00	5.05	6.61
Increase in output (percentage)		405	561
Decrease in unit cost (percentage)*		–46	–51
Low-volume, double shift			
Available hour index	1.00	2.17	2.17
Throughput index	1.00	1.16	1.52
Output index	1.00	2.52	3.30
Increase in output (percentage)		152	230
Decrease in unit cost (percentage)*		30	37

*Assuming that the elasticity of unit cost with respect to output is –0.388, based on econometric analysis found in Miller, "Potential Impacts of Robotics."

its as well, over a 200 percent increase in output could be achieved. Their estimates are also based on current operating times—no changes in shifts per day or days per year.

The potential effects of utilizing nonscheduled production time and of recouping time lost to nonproductive uses during scheduled operations are combined in Table 8. To utilize all of the time normally not scheduled for production—plant shutdowns, holidays, Sundays—the plant would sometimes have to operate with skeleton crews, or even operate unmanned during some periods under the control of computer systems. Thus if we assume that hours available for production could be increased to the upper limit, we should consider the case of robots used in conjunction with other CAM technologies. For the robots-with-CAM case, high-volume machining operations show a potential output increase of 80 percent. Mid-volume manufacturing, and low-volume producers already on a double shift show a potential output increase of 360 percent and 230 percent, respectively. For low-volume producers operating on normal single shifts, the potential increase is 560 percent. In mid- and low-volume manufacturing, potential increases in output are almost all the result of increasing planned production time. In high-volume machining operations, the contribution of increasing throughput per period is somewhat greater than the contribution of increasing hours planned for production.

Assuming that there is a market for additional goods, an increase in output can be viewed as a reduction in unit cost. Estimates of the percentage decrease in unit cost that would result from the percentage increases in output are given in Table 8. These are derived from a study documenting the extent to which unit cost tends to decrease as the level of output increases across metalworking industries.[9] In high-volume manufacturing, unit cost could be decreased by 13 to 20 percent by increasing output by the indicated amounts. Low-volume single-shift and mid-volume producers could reduce unit cost by 40 to 50 percent by increasing output by the indicated amounts.

For the sake of comparison in our discussion of the effects on labor costs earlier in this article, we noted that on average total cost would be reduced by roughly 2 percent if 20 percent of all fabrication workers were replaced, and by roughly 5 percent if 50 percent of all fabrication workers and 25 percent of all assemblers were replaced. Potential cost savings theoretically obtainable from increasing machine utilization appear to be roughly an order of magnitude larger than savings that could be realized by replacing production labor costs under the low- and medium-replacement scenarios. Even if almost all production labor costs were eliminated, the savings that could be realized in low- and mid-volume manufacturing by increasing output would be two to three times greater.

INVENTORY AND WORK-IN-PROCESS COSTS

It has been estimated that in typical batch production operations only 5 percent of the throughput time is required for actual production—a workpiece spends 95 percent of the time in transit or in storage. This suggests that a piece requiring 10 hours of machining time would take at least 8 days and up to several weeks to pass through the factory. Improved work scheduling in combination with higher rates of machine utilization could dramatically reduce the time required to move work through the factory.

Inventories of work in process and finished goods on hand at the end of the year typically comprise between 10 and 30 percent of the value of shipments. Robotics and flexible manufacturing systems will not necessarily decrease the levels of inventory on hand. In fact they may require increased inventory levels, since higher levels of machine utilization will mean higher output rates, and it will be more important to keep a buffer supply of work always available to keep the machines busy.

The financial benefits of reducing the time it takes to move inventory through the plant can be seen from the following example. Suppose a shop with an average of $1 million annual revenues has a 100-day average lead time, which is not atypical. If that lag could be instantly cut to 10 days, the firm would be able to

9. Steven M. Miller, "Potential Impacts of Robotics on Manufacturing Cost within Metalworking Industries" (Ph.D. diss., Carnegie-Mellon University, 1983).

increase a total output for that year by 25 percent and put the next 90 days' revenue, $250,000, into the bank, without increasing costs at all. Deliveries would obviously be speeded up, to the benefit of customers. The added revenue would effectively be converted from unavailable working capital into liquid cash. The firm would also save significantly on out-of-pocket warehousing and other inventory-related costs.

CAPITAL COSTS

In a general purpose batch production facility, capital equipment and labor are shared among a large number of products, since the requirements for a particular product are not large enough to tie up all available equipment. As discussed earlier, the flexibility of a job shop is achieved at the cost of reduced levels of equipment utilization. This can be thought of as capacity loss due to sharing. Increased machine utilization resulting from the adoption of robotics can be viewed as recouping some of the capacity that was lost as a result of capital sharing. In many instances increases in output would cut fixed cost per unit produced drastically, despite the added expenditures for the robots and the accompanying manufacturing systems. The other major effect of programmable automation on capital cost would be a reduction in the long lead times incurred in the sharing context, which would reduce some part of the inventory cost, as already mentioned.

The capital costs of a flexible manufacturing system at present are typically somewhat greater than for conventional types of automation.[10] Currently computer controlled flexible automation is expensive, primarily because of software costs. However, this differential might be reduced to some extent or even eliminated if such systems are built from relatively standardized elements, including programs, as opposed to being custom built, like most specialized production facilities of today.

In general, if capacity could be increased significantly, flexible manufacturing systems could reduce both fixed and variable costs of batch production per unit. Figure 5 shows how robotics and CAM promise to shift the existing average unit-cost curve envelope closer to the ultimate lower limit—materials cost—over much of the spectrum of the production rates, particularly the mid-to-high-volume range.

THE CONSEQUENCES OF DRAMATIC IMPROVEMENTS

Suppose, as we suggest, that unit cost in batch and custom manufacturing could be reduced to below current levels, mostly as a result of expanding the capacity, and output, of existing facilities.[11] There are several consequences

10. However, the flexible manufacturing systems typically have a greater range of capabilities than the equipment it replaces. While overall capital cost might be higher, cost per unit of capability—if this could be measured—might well be lowered for the computer controlled flexible manufacturing system.

11. Much of the savings could in principle be achieved without eliminating labor. However, the higher machine utilization rates can be achieved only by using computers and robots to control the flow of work within the whole factory, eliminating the need for much of the hands-on labor, which in turn eliminates worker-related slow-downs and bottlenecks. If capacity increases are achieved, it would be profitable to pay some of the current workers just to stay out of the way, in order for the machines to be more fully utilized. However, this is unlikely to be the most productive or socially

for employment, depending on how demand for the product changes as price declines. This relationship is given by a parameter referred to as a product's own price elasticity of demand, defined as the percentage increase in demand for a 1 percent decrease in price.[12] A distinction is usually drawn between three cases:

1. Inelastic. A 1 percent reduction in price leads to less than a 1 percent increase in quantity demanded. As a result price cuts decrease total revenue, and probably profits.

2. Unitary. A 1 percent reduction in price leads to a 1 percent increase in quantity demanded. Price cuts leave total revenue and profits unchanged.

3. Elastic. A 1 percent reduction in price leads to more than a 1 percent increase in quantity demanded. Price cuts result in an increase in total revenue and profits.

First, take the case of a relatively modest price cut. Suppose that firms within an industry retrofit robots into existing facilities. They realize a 10 percent decrease in production labor cost, and a 25 percent increase in throughput, at the cost of 5 percent increase in annual capital outlays. Based on these assumptions, labor requirements per unit of output would decrease from 1.0 to 0.72, a 28 percent decrease. Given this decrease in unit labor requirements, demand within this industry would have to increase by 39 percent—1.0/.72-1.0—to keep the same number of workers employed as before. To calculate the reductions in unit cost, which we assume are passed on as price cuts, we figure that the cost proportions in this industry are representative of several industries within the fabricated metals sector (SIC 34).[13] Assuming the improvements and costs mentioned before, this industry would realize a 21 percent reduction in unit cost. The price elasticity necessary to induce enough increase in demand to keep all the displaced people employed is given by

$$\frac{\text{39 percent increase in output}}{\text{21 percent decrease in price per unit}} = 1.85$$

In other words, for each 1 percent decrease in price, there would have to be a corresponding 1.86 percent increase in quantity demanded to generate enough additional employment so that displaced workers could remain within the industry. For the sake of comparison we note that the elasticity for household appliances has been estimated to be roughly 2.0, which makes it one of the most price responsive of all consumer goods. We would not expect capital goods or most other products of the metalworking industry to be as price responsive as consumer appliances.

Unitary elasticity of demand—with the percentage increase in quantity demanded equaling the percentage decrease in price—is the most reasonable assumption for most intermediate goods, lacking other data. Given a price elas-

acceptable use of human resources. It also depends on being able to sell the additional output.

12. The demand for a product depends on its own price, as well as on the price of other products that could be used as substitutes. In the discussion we assume that only the price of the product in question is varying, and that prices of other products and of other important variables, such as income levels, remain constant.

13. We assume the following cost proportions: production labor cost equals 23 percent of value of shipments, nonproduction labor cost equals 7 percent, capital cost equals 20 percent, and materials cost equals 50 percent.

ticity of unity, a 21 percent reduction in price per unit would induce a 21 percent increase in output with no change in revenues. With the reduced unit labor requirements, as before, this outcome would still leave 13 percent of the workers previously employed in the industry displaced.[14] Demand for some consumer tools, including automobiles and appliances, is more dependent on price. If the price elasticity of automobiles were 2, a 21 percent reduction in price or cost would induce a 42 percent increase in sales. Even with the reduced unit labor requirements as given before, this would yield a 2 percent increase in employment requirements and produce additional revenue and profits for the manufacturers as well.

Robot use could easily result in more dramatic reduction, both in unit labor requirements and in prices.[15] Suppose all producers within an industry were to refurbish their existing facilities completely, or even build new plants. For the sake of discussion let us assume that the new facilities realized a 20 percent reduction in production labor cost and achieved a 100 percent increase in output, at the expense of a 20 percent increase in annual capital outlays. In this case unit production cost would drop by 50 percent. With the reduced unit labor requirements, output would have to expand by 250 percent to create enough work to keep all the displaced workers employed in the same industry, which would require a very high price elasticity of 5. If demand increased only by as much as prices decreased, 50 percent, the employment of production workers would drop 40 percent from previous levels.

These examples suggest that if robotic technologies were to be widely used, not all displaced workers could be expected to be reemployed in their current industries, as a result of price reductions and increased demand. The effects of price reductions on demand, and the net employment effect, balancing job displacement and job creation, will vary considerably among the various metalworking industries, depending on the nature of the product and its market. The logical conclusion is that employment of production workers in most manufacturing industries will decrease, despite substantial improvements in productivity within these industries and possible increases in production.

This does not mean, however, that total employment in the economy as a whole would decrease. Substantial reductions in prices of intermediate and capital goods—for example, 20 percent in the first example, and 50 percent in the second—should reduce the cost of manufacturing consumer goods, and of creating new goods and services, both of which will increase the consumer's real buying power. This will in turn stimulate effective demand for other goods and services which should create new employment opportunities.

The foregoing analysis assumed that the only way to utilize the extra capacity made available by robotic systems was to increase the output of the goods that are already produced in that factory or

14. 0.72 labor units/unit of output × 1.2 units of output = 0.87 units of labor = a 13 percent decrease from the base case.

15. As an aside, if demand were to increase by several hundred percent—and in some cases it might—many established organizations would not be prepared to cope with the increased complexity of organizing their business. It would strain the organizational structure, especially information processing capabilities. As a result producers sometimes purposely restrain their technological capabilities in order to keep their business manageable.

industry. However, there is an option of making greater use of the expanded capabilities and of the flexibility of robotic production systems to produce a wider range of products, and to manufacture new, high-performance products. Thus simply looking at the price elasticity of demand for current products might substantially underestimate the extent to which additional flexible capacity could be utilized. If the benefits of robotics and other types of programmable manufacturing technologies are to be fully exploited, there needs to be a concurrent emphasis on the development of new products to utilize the expanded capabilities. A new strategy that places much more emphasis on product performance, and less on standardization and cost reduction, might require an abrupt shift in many existing corporate strategies.

To summarize, the primary economic benefit of robotics is likely to be a reduction in the real cost of manufacturing products made in small-to-medium batches. Capital goods, machine tools, and the other types of durable equipment, as well as the parts used in them, are largely batch produced.[16] Thus the price of capital goods in relation to final products can be expected to decline significantly over the next quarter-century. This will cause secondary ripple effects in the prices of other manufactured goods and services throughout the economy. This in turn will reduce the real price of final output of mass-produced consumer goods, as well as the real price of output of the nonmanufacturing sectors. Final demand would also be stimulated to some extent, depending on the sensitivity of final demand to price. For consumer goods, high price elasticities tend to be more the rule than the exception.

Lower production costs will also have a beneficial impact on the rate of inflation. Insofar as inflation is caused by too much money chasing too few goods, an increase in productivity is perhaps the best way to break out of the vicious cycle. Ultimately, such changes will also affect other important macroeconomic variables, including the overall level and composition of employment, and the level of distribution of income. These second-order effects, while less immediate, may have greater ultimate importance than the immediate improvements in labor productivity in manufacturing. It is beyond our present scope to attempt to forecast the detailed nature, magnitude, or time phasing of these broad economic impacts.

We expect that improved robots and substantial reductions in the price of intermediate and capital goods will play an important role in facilitating the development of several capital-intensive growth sectors in the economy, including hazardous waste management, biotechnology, undersea mineral extraction, and space manufacturing. These sectors would also provide employment. It is important to know whether or not the levels of economic growth required to absorb workers displaced by robotics and other technological changes can be achieved in the economy as now structured. If the required levels of economic growth, and employment, cannot be achieved as a result of cost saving improvements in the manufacturing process, resources may have to be reallocated to encourage the creation of new

16. The major exception is automobiles. Large appliances, such as refrigerators, air conditioners, and washing machines, are also mass produced durable goods, but since they are sold to consumers they are not classified as producers' durable equipment.

products or services, or the development of new frontiers, such as the oceans and space. This would require a reevaluation of traditional policies of stimulating economic growth by encouraging aggregate investment.

EMPLOYMENT EFFECTS

The limited experience to date is consistent with the point of view that for the overall economy industrial robots pose little serious threat to employment in the coming decade. Assuming 1600 robots in use in the United States in 1977 and 6800 by the end of 1982, and assuming that each robot displaces two or three workers on average, robots may have displaced 3000 to 5000 workers by 1977 and up to 20,000 by the end of 1982. The last figure cited would represent about one-fifth of 1 percent of the approximately 9.7 million semiskilled operatives and unskilled laborers employed in manufacturing industries in the United States in 1980. From this perspective the effects of employment up to now have been negligible.

But extrapolating the experience of the recent past into the future overlooks the concentration of effects in a relatively narrow sectoral, occupational, and regional setting. In particular, robotization along with other and frequently related developments could diminish employment opportunities for semiskilled operatives and unskilled laborers in durable goods industries, especially in the metalworking sector. These considerations imply that the displacement will be of sufficient magnitude, at least in some industries and regions and for some groups of workers, to be a cause for concern.

Robots and CAM continue the long-term trend toward mechanization and computerization that have made for relatively static employment in manufacturing industries in the United States, despite increased output. The significance of robotization in the next decade or two is that it will add to the combination of factors that have retarded growth in manufacturing employment. The percentage of the work force employed in industry—mining, manufacturing, and construction—has steadily declined from about one-third of the work force in 1959 to barely one-quarter in 1977. Projections by the Bureau of Labor Statistics indicate that this trend will continue. In addition the service content of each major sector of the economy, as represented by the proportion of the industry work force classified as nonproduction workers, has steadily increased. In 1980 over one-quarter of the employees in manufacturing and mining were performing managerial, professional, clerical, sales, or supervisory activities.

In 1980 about 14 million people were employed as production workers in manufacturing. Nearly half of all production workers, and nearly half of all manufacturing workers, are concentrated in the metalworking industries. Manufacturing production workers are classified into three main categories: craft—skilled workers; operative—semiskilled workers; and laborers—unskilled workers. Most of the routine repetitive jobs that currently lend themselves to automation and robotization are performed by these semiskilled nontransport operatives, who comprise nearly 40 percent of total manufacturing employment. In several job categories that appear most amenable to robotization, almost all employment is concentrated in metalworking. For instance, nearly all the 1 million operatives of metal cutting and metal forming machines are employed in these industries.

While the majority of jobs that can be robotized are semiskilled operative jobs, there are already robot applications in heat treating, sheet metal work, and forge and hammer operations, which are classed as skilled jobs. As computer aided design and manufacturing become more integrated, and if factories are to exploit fully robotics and other types of programmable automation, a larger fraction of the so-called skilled metalworking crafts will be vulnerable to automation.

The technical potential for replacing workers by robots has been estimated from an analysis of industry employment by occupation and from survey responses to the potential for substitution within a given occupation. Two levels of robot technology are distinguished: robots similar to those on the market in 1981—Level I; and robots with rudimentary sensory capabilities—Level II. In 1980 there were nearly 6.7 million production workers employed in the metalworking sector in the United States. Of these, nearly 5 million worked within the three broad categories of jobs most amenable to robotization—metalworking craft workers, semiskilled machine operators, and laborers. Based on the survey results, we estimate that Level I robots could theoretically replace 16 percent of the workers in these three groups, and that Level II robots could theoretically replace 40 percent of the same population of workers. Thus, if all the potential for job displacement of Level I robots were realized in metalworking, more than 800,000 jobs could be eliminated. If Level II robots were available and fully exploited, an additional 1.2 million jobs, or a total of nearly 2 million jobs, could theoretically be eliminated.

Extrapolating the data for metalworking to similar tasks in other manufacturing sectors, it appears that Level I robots could theoretically replace about 1.5 million metalworking craft workers, semiskilled machine operators, and laborers, and Level II robots could theoretically replace about 4 million of the current 10.4 million of these workers. However, the time frame for this displacement is at least 20 years. In the coming decades the capabilities of robots can be expected to increase further, and their potential for displacing operatives will increase. On the other hand, not all of the potential displacement will actually be realized, and if the economy grows as anticipated, some of the jobs lost due to displacement of people by robots could be offset by increases in other manufacturing employment.

By around 2010 it is conceivable that more sophisticated—Level III?—robots will replace almost all operative jobs in manufacturing, about 9 percent of today's work force, as well as a number of skilled manufacturing jobs and routine nonmanufacturing jobs. Concerted efforts should be made by the private and public sector to prepare the future work force to respond to these changes. Even though several million jobs in the current manufacturing work force are vulnerable to robotization, the transition seems hardly catastrophic on a national scale, provided that new job entrants are properly trained and directed. In our view the oncoming transition will probably be less dramatic than the impact of office automation over the same period. By 2010 most current operatives would have retired or left their jobs. The jobs would not disappear all at once, and robot manufacturing programming and maintenance will provide some new jobs, although we think most new jobs will not be in manufacturing, despite the rapid growth of the robotics industry. New growth sectors in the economy, including undersea and space exploration, may also provide many new jobs. The important conclusion is that young people seeking jobs in the near future will have to learn marketable skills other than welding, machining, and other operative tasks that are now being robotized.

Even though the adjustment problems seem manageable, the potential for social unrest in specific locations cannot be dismissed so lightly. Consider the following points:

1. Nearly half of all the unskilled and semiskilled operative workers—the types of jobs that could be replaced by robots—are concentrated in four metalworking industries (SICs 34-37). Almost one-half of all production workers in these four industries are geographically concentrated in the five Great Lakes states—Indiana, Illinois, Michigan, Ohio, and Wisconsin—plus New York and California. In these first five states the metalworking sector also accounts for a large percentage of the total statewide employment in manufacturing. Adjustments in response to the rapid diffusion of robotics may be intensified in these areas. The adverse effect of not improving the productivity and competitive standing of these industries would also be concentrated in the same few states, of course.

2. Older established workers will generally be protected by union seniority rules, except in cases in which the whole plant closes. Unfortunately this is happening with increasing frequency. Even in the newest, most efficient plants, some younger workers with less seniority may be bumped. In either case the displaced worker starts again at the bottom of the ladder. Thus reemployed displacees are also more vulnerable to subsequent layoffs. A class of perpetually insecure, marginal workers could result. This would be a potential source of social problems and political dissension.

3. The states where jobs are most likely to be lost to robots are mainly in the north central region, where industry is also most unionized, plants are oldest, and wages are highest. The Sunbelt states, where many new jobs are being created, have new plants and lower wages. Many displacees will have to migrate to other regions. Those unable to upgrade their skills sufficiently might have to accept lower-paying service jobs or join the underclass of insecure marginal workers who never became established with a stable employer.

4. There would likely be a disproportionate effect on racial minorities and women. Nonwhites account for only 22 percent of the national work force but over 16 percent of total employment in semiskilled and unskilled manufacturing jobs. Women employed in semiskilled and unskilled manufacturing jobs are less likely to be represented by labor organizations than their male counterparts. De facto economic discrimination will accordingly increase.

5. Unions representing the affected categories of workers will probably ex-

perience sharp declines in membership and political and economic clout. A policy of organized resistance to the introduction of labor-saving technologies might seem attractive to fearful workers and their unions, resulting in a severe drag on the productivity of the manufacturing sector.

JOB OPENINGS

In the occupations expected to be primarily affected by robotization, the job openings likely to be created by attrition in the 1980s provide a basis for assessing policies dealing with displacement during the next two decades. Attrition rates for semiskilled workers in metalworking are approximately 3 percent per year, depending on the sex and age distribution of the persons employed. However, these figures substantially underestimate the number of people transferring out of specific occupations, since they only include people who leave the establishment.[17] A 3 percent annual attrition rate suggests an annual average of about 170,000 job openings among blue-collar workers in the metalworking industries alone during the 1980s.[18] If one-half of these openings could be filled by operatives displaced from other jobs by robots, an average of over 85,000 jobs a year, or nearly 850,000 during the decade, could be filled while still accommodating some first-time job seekers.

As the number of new entrants into the labor force declines in the 1980s because of the drop in birth rates after the mid-1960s, older and more experienced blue-collar job seekers will face less competition from younger workers than in the 1970s. However, an unresolved question at this point is the extent to which economic growth or continued recession in basic industries such as automobile and steel will increase the number of job seekers competing with employees displaced by robotization. In a declining industry, moreover, openings that would otherwise be created by attrition are often left unfilled. Turnover openings that would arise from occupational mobility can fall off sharply as fewer persons are added to the employment roll and few quit voluntarily to take any other positions.

Many of the blue-collar workers displaced by robots would have the educational qualifications for more skilled training, which would lead to a better-paid position in other occupations. The traditional stereotype of a factory operative has been that of a person with limited education, often a functional illiterate. For a generation the typical educational level for operatives was in fact below that of the overall work force. In the mid-1950s, for instance, the median number of years of schooling completed by operatives was 9.5, compared with 11.7 years for all employed civilian workers. By 1978, however, the median for operatives, excluding transport operatives, was 12.1 years, compared with an overall work force average of 12.6 years. Operatives as a group tend to have at least a high school education or its equivalent. This can provide a basis

17. Attrition is used to refer to workers who leave the establishment as a result of quits, discharges, permanent disability, death, retirement, or transfers to other companies. The other main source of labor turnover is layoffs, suspensions without pay for more than seven consecutive days, initiated by the employer. Together the attrition and layoff rates comprise the total separation rate.

18. Assume 6.7 million blue-collar workers in metalworking jobs, and an average rate of 3 percent decrease per year over a 10-year period.

for further specific vocational training or for further higher education in a two- or four-year college.

Changes in national priorities could also expand the range of job openings and thus make obsolete projections such as those of the Bureau of Labor Statistics, which are based on the experience of the recent past. A shift in national priorities favoring more adequate home care, income support, and medical services for the elderly, the retarded, and the handicapped would be reflected in new kind of jobs for persons with the appropriate training. Private and public efforts to rehabilitate physical infrastructure—bridges, subways, water and sewer systems—could create a large number of job openings that could partially be filled by displaced production workers.

The 40-hour standard work week has remained unchanged in most manufacturing industries for the past 20 years. White-collar workers typically enjoy a shorter work week. For example, two-thirds of all office workers in the finance, insurance, and real estate sectors now work less than 40 hours per week. A gradual reduction in the standard work week, leaving hourly wages unchanged, would diminish job losses by spreading the available work over more employees. Clearly the reduction in annual work hours could be accomplished in various ways. Sabbaticals, now confined to teachers and to some civil servants and steelworkers, could be extended generally to production workers. Required sabbaticals at part-pay could be used to explore another occupation, to care for babies, or to become a student again. Blue-collar workers returning to or first entering a university while on sabbatical could provide a new market for the services of colleges and universities faced with shrinking enrollments because of low birth rates two decades earlier.
explore another occupation, to care for babies, or to become a student again. Blue-collar workers returning to or first entering a university while on sabbatical could provide a new market for the services of colleges and universities faced with shrinking enrollments because of low birth rates two decades earlier.

CONCLUSION

Despite the present limitations on robot capabilities and the incremental penetration of robotics into the economy, over the next two decades it would be wrong to assume that this phenomenon is just a continuation of mechanization and automation. That would be only partly true, for two reasons.

First, as extensions of human physical abilities, earlier machines took over many physical tasks, but in no sense did they exclude humans from the production process. Humans have always been needed to operate the machine. What is new is that sensate robots are now becoming capable of replacing humans. Thus robots are now or soon will be competing directly with humans for certain routine jobs involving relatively low levels of visual or tactile ability and relatively little intelligence—that is, problem-solving ability. The implications for future employment opportunities and education and training needs are profound.

Second, while robots are having their biggest immediate economic effect in the manufacturing sector, their greatest long-term effect is very likely to be in other areas. Four examples of future applications come immediately to mind: (1) household robot servants; (2) robot assistants for the physically handicapped, including leg amputees, paraplegics, quadriplegics, and so on; (3) robot soldiers; and (4) robot workers for deep underground mines, underwater activities, and industrialization of outer space. By the middle of the twenty-first century historians looking back may see much greater significance in one or more of these advanced applications than in the more pedestrian ones we have discussed at length in this article. Nevertheless the first steps must be taken successfully before any subsequent developments can occur. For the rest of the twentieth century the place of robots will be mostly in the factory.

ANNALS, *AAPSS*, **470**, November 1983

Technology for the Factory of the Future

By JAMES LAWSON, JOHN REIDY, G. FREDERICK RENNER, KAREN ROSEN, and LISE SMOLAK

ABSTRACT: It is becoming increasingly apparent that major investments in technology will be required in the near term if industrial firms are to maintain their competitive edge in the marketplace. This development will be both driven and limited by the state of the art of technologies associated with the evolving factory of the future. An examination of technological developments and emerging applications provides a better understanding of the factory of the future and its implications for the existing social and economic order.

James Lawson is a projects manager in the Applied and Technical Economics Department of Battelle's Columbus Laboratories. His research has focused on technological effects on economic growth and socioeconomic systems. John Reidy is a research engineer in the Equipment Development Section of Battelle and is responsible for robotic programs. His research focus is creative mechanical design for special purpose equipment. G. Frederick Renner is a research scientist in the Intelligent Device and Microcomputer Systems Group. His research focus is intelligent control systems and robotics. Karen Rosen is a research scientist in the Manufacturing Engineering, Components, and Safety Section. Her research has focused on manufacturing productivity improvement and the integration of robotics and related systems in the manufacturing environment. Lise Smolak is a researcher in the Applied and Technical Economics Department. Her research has focused on technological considerations at the market, industry, and firm levels.

THE advances to be made in the factory of the future are intrinsically linked to technology and technological development. It is therefore important to have an appreciation and general understanding of current technologies, fringe technologies, research and development activities, applications, and the relation between application and technological development. This article provides a review of technology, applications, and their interrelationships.

The primary focus of the article is robotics and related technologies. Although the technologies important to the evolution of the factory of the future will go well beyond that of robotics, robotics will be the near-term technological focus in the development of the factory of the future.

CURRENT TECHNOLOGIES

The term "robot" has been applied to many different devices and systems without any general agreement as to its meaning. The Robot Institute of America defines a robot as "a programmable, multi-function manipulator designed to move material, parts, tools, or specialized devices through variable programmed motions for the performance of a variety of tasks."[1] This definition applies most directly to the devices commonly referred to as industrial robots, and it is these devices as a class that will primarily be involved in the development of the factory of the future.

Current industrial robot technology has evolved from various precursor technologies such as teleoperators, numerical control machine tools, and material handling. This evolution has been fairly recent and is reflected in the structure of current products, many of which are built directly of subassemblies from the precursor technologies. There are also, however, new and sophisticated robots that reflect the intensive development the field has experienced over just the last decade.

The rapid development of the technology has produced a market characterized by extreme product variety. Product configurations, sizes, capabilities, and costs range so widely that the term "industrial robot" is no more specific than the term "camera." Contributing to as well as preserving the variety of product offerings are the numerous situations and tasks for which industrial robots are designed and utilized.

For purposes of discussion, a robot can be considered to consist of four major subsystems: a manipulator, an end effector, a control system, and optional sensor systems.

The manipulator

Manipulator technology has evolved largely from teleoperated systems, systems in which a manipulator is under the direct control of a human operator. Teleoperated manipulators were developed originally for such work as the handling of hazardous nuclear materials. Most were master-slave systems in which a master arm, manipulated by the operator, was mechanically coupled through a wall to a slave arm that performed the desired operations without hazard to the operator; cables and other linkages transferred the operator's muscle power directly to the slave arm. Newer, advanced manipulators, powered by hydraulic or electric actuators, eliminate the need to couple the master and slave arms mechanically, allowing an increase in the distance between the operator and

1. Joseph P. Engelberger, *Robotics in Practice* (New York: American Management Association, 1980), p. 8.

slave arm, but also requiring communication links to transfer the control information. Through the feedback of positioning information for each actuator, these manipulators can be positioned at any point throughout their movement with a high degree of accuracy. Such manipulators are used in servo robots, which perform most of the complex operations accomplished robotically in modern factories, such as spot welding and spray painting.

Robot manipulators are designed in response to a variety of functional requirements, many of which are at odds with each other. Major functional criteria, for example, require manipulators capable of lifting significant weight over varying distances; these requirements of strength and reach inherently oppose other requirements for positional accuracy and speed. In addition, the larger a robot arm must be to hold and extend a weight, the more power will be required to move the arm at an acceptable speed. As a result, in certain applications a robot can perform the same task as a human operator but with a power requirement that may be 50 times greater.

End effectors

The end effector of a robot is roughly analogous to the human hand. In effect the purpose of a manipulator is to position and orient an end effector at the proper time for it to perform its function. For many robots this function is to grasp or release a workpiece. This is accomplished by a variety of mechanisms, depending on the type of material and the shape of the workpiece. Most end effectors grasp or clamp the workpiece mechanically, although some use special devices such as suction cup holders, or special tools adapted to robotic use, such as a welding gun or a paint sprayer.

As an indication of the variety of types of end effectors required for individual applications, most robots are not provided with an end effector at the time of purchase. Rather end effectors are usually designed for the specific applications for which they are intended.

Sensor systems

In general the use of external or task-related sensors is the least-developed major aspect of industrial robotics. The simplest and most commonly used sensors are proximity devices that detect the presence or absence of a workpiece. Used with appropriate programming, these sensors provide a significant degree of flexibility in tasks such as removing assorted items from stacked pallets. With the addition of sensors to determine grip position, the ability to determine the size of the object being grasped is also possible.

A significant increase in both technological sophistication and cost occurs in moving to the other generally available type of general purpose sensors. These are the computer vision systems, varieties of which are available from perhaps a dozen suppliers. The term "vision" exaggerates the primitive shape recognition capabilities of most commercial systems, but these systems can provide a significant amount of information regarding a task and a situation. Such information is needed for a robot to adapt to both planned and unplanned variations. Programming to use and act on this information must also be provided, however, and this requirement is the major bottleneck in the movement of computer vision technology into the factory.

Other sensor technologies have been developed to assist in specific tasks—particularly arc welding, for which systems are available that use electrical through-the-arc and other sensing techniques to provide information about the welding process. Based on the derived information, a computer control system adjusts welding parameters to compensate for variations and to produce a consistently good weld.

The controller

The arm of a robot needs a brain to control it. In modern industrial robots this function is performed by an electronic computer. Robot manufacturers customarily provide programs in their computer for two important functions. First, a method is provided to teach the robot a series of operations. Second, programs in the controller perform all of the closed-loop control calculations that cause the manipulator to perform the user-supplied operation sequence.

Two basic approaches are used in teaching robots to perform their jobs. In the most common approach, called walkthrough or leadthrough teaching, the user moves the manipulator through the desired operations either by using a push-button teaching pendant or by physically moving the manipulator itself. In the second approach, called off-line programming, a specialized language is used to describe the required sequence of operations. This description is converted by computer into a form the controller can use. An advantage of off-line programming is that a robot can be working while new programs are being written for it. A disadvantage, in addition to limited availability, is that off-line programming requires skills similar to those required for computer programming. In contrast, walkthrough teaching can be performed by shop floor personnel.

A robot program, regardless of its method of generation, consists of a list of manipulator positions. In addition some measure of speed may be recorded for tasks, such as spray painting, in which speed is important. Using this information, plus feedback signals from position sensors on the manipulator joints, the controller causes the manipulator to move smoothly and accurately along a desired path. In some robots special provisions allow minor variations in the path, depending on specific decision criteria.

There are of course limitations in the capabilities of current controllers, chief of which is an inability to accommodate variations in task or environment. Significant advances in technology will be required before this capability, assumed in human workers, can be provided in robots.

FRINGE TECHNOLOGIES

Among the pertinent fringe technologies are artificial intelligence, microcomputer technology, and military systems.

Artificial intelligence

There is surprisingly little interchange between the fields of artificial intelligence and industrial robotics, despite their obvious logical connection and their connection in academic research. Current commercial applications of artificial intelligence are concentrated in expert systems, which emulate human symbolic problem solving. Expert-system development involves the compilation, through extensive interviews, of relevant facts, procedures, rules-ofthumb, and other knowledge held by human experts; this information is stored in an arrangement similar to a computer data base, where it is accessed by a computer program that sifts through the information and develops assertions from it. Unlike the interest in expert systems, very little attention is being directed toward developing capabilities analogous to human motor skills. Sensory skills are largely represented by the vision work discussed earlier.

Microcomputer technology

The dramatic increase in capabilities and drop in price of digital electronics has been well publicized. Although there is a delay of several years between the development of a new processor and its incorporation into a robotic product, microcomputer technology is having an increasing effect on both the capabilities and the availability of better robot controllers. Specialized processing tasks such as those encountered in computer vision are a topic of current research into how they can best be accomplished with very large-scale integration (VLSI) technology. These studies are considering architectures—such as simultaneous multitasking—radically different from the linear vonNeuman computers that have formed the basis of the computer revolution over the past 25 years.

Military systems

The military has been actively involved in the development of systems and technologies pertinent to the factory of the future. Some of these efforts, described most appropriately as manufacturing technology, involve the robotic manufacture of systems and components directly for the military. The military is also developing systems for use in military field applications that, although not directly pertinent to the factory of the future, are appropriately considered fringe technologies. These systems include a variety of remote sensors and the associated artificial intelligence, and controls required for data interpretation and in some cases decision making. The military has also been involved in the development and use of a variety of teleoperated systems for such applications as the remote handling of hazardous materials. In addition remotely piloted air and ground vehicles have been developed, and the navy is currently developing several vehicles that will be capable of autonomous operation underwater.

RESEARCH AND DEVELOPMENT ACTIVITIES

Current research and development is addressing the problems of manipulators, end effectors, sensors, and robotic systems.

Manipulators

For manipulators powered by permanent magnet (PM) DC electric motors, advances in rare-earth magnet materials such as samarium-cobalt may produce significant benefits. Rare-earth PM motors are capable of providing torques high enough to allow direct-drive configurations. If the mass of the gear trains used now with the PM motors can be removed, manipulators can be made lighter and faster.

In an attempt to move current robots at required speeds and minimum deflections, research is being conducted on the control of flexible manipulators. In these systems the structures will be designed to allow for deflection; the deflection will be accounted for and compensated for by a control system, probably based on sensor input. Such systems will provide the reach, payload, and sphere of response of current manipulators, but without the large power requirements that are due to the size of the current machines.

End effectors

A major research effort in this area is directed at facilitating parts assembly through the incorporation of compliance in the end effector. This work is intended to result in the use of the part and the mating piece themselves to assist in the guidance of the two parts. An example would be an interactive adjustment between the hand of the robot and a part such as a screw, such that finely threaded fasteners might be used without stripping thc threads of the fasteners.

Japan is the leader in research into the use of articulated end effectors or hands for robots. Many such Japanese devices are direct models of the human hand, with cables performing the function of tendons; this allows the relatively bulky actuators to be located away from the hand itself. Stanford University has also conducted studies on a simplified three-finger hand design. Extensive development of this technology will be required before such devices offer all of the capabilities required for general assembly work. Full duplication of human hand functions is not likely in the near future.

Sensors

The topic of vision has largely dominated sensor research related to robotics. Sensor capabilities now being pursued include utilization of three-dimensional gray scale, and color information. Performance improvements are also being sought in speed and resolution. One recent product announcement, for example, claims a capability for identifying and locating a single part within a bin of jumbled parts.

Tactile sensing is also receiving attention. There are several methods of tactile sensing. One involves multiple-axis force sensing in the wrist; this provides kinematic information such as might be required in assembly operations. Another type of tactile sensing involves a two-dimensional image of pressure levels; this bears a resemblance to vision and indeed is often suggested as a replacement for vision in certain applications.

Closely integrated within sensor research is robot programming that incorporates sensor information. Clearly a sophisticated sensor cannot improve task performance significantly unless the controller can both acquire the sensor information and make use of it. The development of such capabilities is progressing slowly as improvements are made in the design of robot programming language systems.

Systems

Efforts are currently under way to incorporate mobility into robotic systems, particularly in factory robots. With such mobility robots could perform operations at more than one work station. The problems of self-location both at the work station and in transience—if robot transportation is to be autonomous—are formidable, but the benefits to the factory of the future are certain to be significant.

Various government agencies are now studying the development of robotic manufacturing systems. One such system is an outgrowth of the air force's integrated computer aided manufacturing program, in which robotic cells are being developed for the repair of turbine blades. The army is sponsoring a similar program exploring the potential for automatically manufacturing electronic components. Other agencies, such as the National Bureau of Standards, are also conducting programs that may contribute to this field.

CURRENT APPLICATIONS

At present only a small percentage of tasks performed in the factory environment are well suited for robotic applications. Commercially available robots cannot necessarily perform all tasks of human operator capability, such as taking parts randomly out of a tote box. Robots generally require an orderly workplace in which parts are delivered to the same location in the same orientation every time.

While these limitations seem stringent, there are many applications for which robots have proven to be cost-effective. In applications thus far a reliability of over 90 percent is generally being realized with productivity gains ranging from 10 to 90 percent. The average cost of operation for a standard industrial robot is remaining relatively constant at $5 per hour.

The major industrial sectors now using robots are motor vehicles, aerospace, metal working, machinery, glass, rubber, and electrical/electronics, with the automotive industry being the dominant customer of robot vendors.

Current applications for robots generally involve highly repetitive tasks of low complexity, and the removal of workers from hazardous environments. The most common functions thus include welding, painting, assembly, and material handling.

Welding

Welding operations account for over 50 percent of all robot installations in U.S. plants, the majority of which are performing automobile spot welding operations. There are several reasons welding has become a very popular robotics application. First, welding is labor- and cost-intensive with labor accounting for 86 percent of welding costs. Second, there is a shortage of skilled welders. Last are the government efforts to remove workers from the welding environment for health and safety reasons.

Material handling

Material handling operations include both transfer of parts from machine to machine and loading of parts into machine fixtures. Material handling is also a highly roboticized operation in industry. Examples of this application include placing and removing pieces from conveyors, loading and unloading pallets, and handling tools and dies. Advantages of robots in materials handling applications include the ability to handle heavy loads, reduction of part damage attributable to improper handling by human operators, and the ability to handle parts that would be hazardous for human operators.

Painting

The ability of some robots to follow complex movements repeatedly makes them well suited for spraying paint. A continuous path-type robot is required for these applications. The advantages of roboticized paint spraying are consistent quality, material savings, and the removal of humans from potentially hazardous environments.

Assembly

Automated assembly is a relatively new application and function for robots. Assembly operations have strict requirements, including high accuracy, speed, sensing capabilities, and complex feedback control systems. The major advantages of robots in this application are higher production rates, lower defect rates, and reduced obsolescence of the assembly line.

EMERGING APPLICATIONS

In the present factory environment, a large percentage of the work day is lost because of scheduling problems, unclear communication of assignments, improper staffing, and poor discipline in the labor and management ranks. These conditions are not easily reversed and provide further incentives for increased use of robotics.

The U.S. robotics market is expected to grow at a rate of 35 percent per year throughout the 1980s. Arc welding is expected to realize the greatest growth over the next few years, with automated assembly becoming the most important robotics application in the mid-to-late 1980s.

The present generation of robots is limited to repetitive tasks of low complexity. As more sophisticated, smarter, and dexterous robots are developed, however, the use of robots for increasingly complex applications will undoubtedly grow. The development of more sophisticated robots requires advances in sensing technologies—such as tactile and vision systems—artificial intelligence, and methods for programming the robot's movement. Growth of emerging applications is dependent upon the transfer of this research and development in advanced and fringe technologies to the marketplace.

Arc welding

It is estimated that a majority of future arc welding applications will require some type of complex sensory perception. Vision sensing for seam tracking is needed before the full potential for arc welding can be realized.

Machine loading

While machine loading is a simple task for most commercial robots, it is not cost-effective in applications that require the robot to sit idle between load/unload cycles. Improved flexibility and mobility in robot systems that will allow multiple functions to be performed in different locations by the same robot will increase the cost-effectiveness of machine loading applications.

Inspection

With the proper programming and sensing abilities, a robot can provide instantaneous, in-process inspection. A variety of workpiece characteristics including size, weight, and actual performance could be tested.

Assembly

Robots are now limited in assembly applications by constraints in sensing, manipulation, and programming. End effectors having the coordination of the human hand combined with vision will enable robots to perform tasks such as picking up and moving randomly located parts, recognizing specific part types mixed in with other parts, performing measurements and related inspection tasks, and conducting difficult assembly tasks in an adaptive fashion.

SUMMARY

Current practice for programming movement is difficult and inflexible and does not lend itself well to these complex tasks. The development of advanced high-level robotic languages will allow humans to give robots specific commands or goals rather than a detailed program of how to accomplish these commands, The robot controller will ultimately utilize artificial intelligence concepts to interpret and carry out these commands. The robot controller will languages need not be intimidating to the user and will employ man/machine interface techniques such as voice recognition, which allows the robot to respond to oral communication.

Integration of robots into emerging applications and total process automation will demand advanced robot capabilities and related changes in process and workpiece design within the entire factory environment.

Integration of robotics into the factory environment

Most factory environments have been designed for operation by human workers, not robots. Straight line setups of machines are often used to allow efficient work flow, where human workers have the flexibility to move and adapt to such arrangements. Robots have generally been designed as stationary machines, in which their reach provides a stationary work envelope, and are therefore less adaptable than human workers within the factory environment. For optimum utilization of robots, machines have been arranged to fit the requirements of the robot. The adaptation of the workplace to the robot's requirements, however, adds substantially to the installation costs of robots and has led to efforts to adapt robots to conventional machine layouts. Additional efforts are being made to integrate the robot more effectively into the human environment of the factory at all levels, from upper management to factory workers.

When robots were first introduced their capabilities were poorly understood, and it was thus difficult to select appropriate tasks. Many people picked and are still picking what they feel are good robotic tasks, only to find out later that the cycle speed or manipulations required cannot be achieved with a robot. Conversely, good applications have gone unnoticed because of a lack of knowledge of the full range of robot skills and abilities. Presently, however, both users and manufacturers are gaining a better understanding of robot capabilities and applications from increased coverage in the popular press, and this is beginning to affect the integration of robots into the factory environment.

Corporate management demands a valid cost justification for such major expenditures as robot installations. As a result economic considerations, such as labor, materials, productivity, energy, accessory equipment, launching costs, and tax credit/depreciation, are important evaluation criteria for robotics purchase decisions. These considerations allow a more comprehensive and realistic approach toward quantifying specific values accrued through increased productivity and throughput, improved inventory turnover, simplified safety and environmental requirements, and reductions in rejects and rework, accidents and compensation costs, and personnel turnover. Some general characteristics of production situations in which an industrial robot is likely to be economical and practical are hazardous environments, repetitive tasks, difficult handling tasks, and multishift operations.

It should be noted that robots improve productivity not so much by working faster than humans, as by their constancy of pace and consistency of operation. Operating consistency provides better part quality and reduced scrap. The integration of a robot into the factory environment involves several considerations:

—need for complete knowledge of the area, process, variations, and production where the robot is to be located;

—reasonable understanding of a robot's capabilities and limitations, and a realization that current-generation robots require a clearly defined orderly workplace oriented toward the robot's requirements;

—importance of maintenance and repair, the effect of malfunctions on the production process;

—education of personnel at all levels who will interact with robots.

Educating the factory worker about robotics enhances the identification of other possible robotic applications. No one knows the idiosyncrasies of the factory environment better than the machine operators. When a possible robotic application has been identified, it is important that the operators be consulted and that these idiosyncrasies be identified. For example, an experienced operator may know that debris on parts causes an assembly line to jam during a pressure-fit task and that he should rub each part on his shirt before he assembles them. The debris problem may rule out the possible use of robotics in this application, and an ineffective application might result from not talking first with the machine operator.

Consultation with all workers is important, since they may be concerned about losing their jobs to robots. If not properly consulted they may be suspicious and antagonistic about the integration of robotics into their workplace. It is important to keep workers informed of changes well ahead of time, to address their fears regarding job security, and to provide retraining if necessary. To ease integration it is desirable to let workers know that they will be retrained in robotics or some equivalent job function. Some firms have found it beneficial to appeal to the pride and competitive spirit of their workers by allowing only the best of the workers to be retrained for robotic-related job functions.

Production supervisors often view new robotics applications as a potential cause of additional problems. A smoother integration of robots into the factory can help them and their workers. It is especially important that the production supervisor understand the capabilities of robots, their advantages and limitations, and have management support in resolving problems caused by their introduction.

Middle and upper managers often regard the introduction of robots as a substantial risk. Middle management should be educated about robots and assured of upper-management interest and support. Upper management needs to be informed of the economic potential of robots for the company, their advantages and limitations, and the potential for problems that may be encountered with robotic applications.

Normally the engineering staff will have the technical background necessary to understand the technologies associated with robotics at the start of any project. In order to ease the integration of robots with this group of personnel, it is often desirable to maintain an information library on robot technology and to send staff members to seminars on robotic applications. When robots are first introduced it is often beneficial to hire experienced engineering consultants to work with the factory engineers in the initial installations. The engineering departments should also consider hiring persons who have experience in utilizing robots and associated technology.

These guidelines can ensure a more successful transition from the current factory to the automated factory.

Technology, applications, and societal relationships

Structural relationships between technological developments and applications are important to the issue of automation in the workplace. In the United States there is a general pattern of structural relationships between a significant portion of our domestic technology development, applications, and society. This pattern evolved from the concept of the military-industrial complex and can be characterized as follows:

1. Government-funded, programmatically focused research and development and introductions of fringe technologies for military, space, and related applications. This is characterized by very high-cost and limited-access technology.

2. Transfer of a maturing fringe technology to limited industrial applications. This is characterized by very high-cost, limited-profitability technology that is suited to a few very technically competent firms.

3. Maturation of the base technology within the industrial sector. This is characterized by steady reductions in cost and improved utilization within the industrial and commercial sector.

4. Technical advances and development of synergistic technologies. In this case "synergistic technologies" refers to technologies that are not necessarily linked to a base technology, but with advancement beyond some threshold level they become synergistic to the base technology. An example would be the relationship of intelligent computers with the current robotic technology base.

5. Emergence of new applications resulting from advances in synergistic technologies. A current example is the development of low-cost microelectronic technology and its introduction to computer base technology for home use. This stage is normally characterized by a large-scale increase in the market of the base technology.

6. Influence of new applications on new technology development initiatives. Using the previous example, home computers influence the development of entirely new software and related technologies.

7. Impact of new applications upon economic and social structural relationships. A base technology normally has a limited societal impact until it is introduced on a large scale. Again using the example of the home computer, the average person's interaction with computer base technology was limited and little change was noticed in the economic and social structure until the

introduction of the home computer. The overwhelming presence of computer base technology with the introduction of the home computer, however, is beginning to change our approach to education, the way we conduct our personal business, and our communications and interaction with others.

8. Economic and societal response and adjustment. This stage can be characterized by changes in market structures; societal institutions come into being while others decay and reposition.

Although robotics is the current base technology associated with the factory environment, there are several emerging synergistic technologies that will foster the large-scale automation anticipated for the factory of the future. These include microelectronics, sensors, intelligent computers, and computer integration of discrete business functions.

The field of microelectronics is providing improved speed and capacity in increasingly small spaces. Microprocessor chips the size of a fingernail can be constructed with the equivalent of 1 million transistors. Current research hopes to expand this to the equivalent of 5 million transistors within the same space.

Although most of the robotics sensor focus is vision oriented, other nonvisual sensor research may have more profound impact in the longer term. Current research and capabilities include rapid determination of atomic structures, discrete structure imaging capability, and thermal and ultrasonic imaging. Although many feel that sensor research and development for robots has been lagging, the total sensor research effort in the United States is currently exceeding $500 million per year.

The development of a truly intelligent computer will have a significant effect upon the factory of the future. The recent commitment of the Japanese to development of the fifth-generation computer may provide the impetus needed for concerted effort to develop functionally intelligent computers.

Technological advances are increasing the opportunity for effectively integrating many discrete business functions into a computer controlled system. Such an integrated system would go well beyond the factory environment and could include a computer integrated system for purchasing, production, inventory, distribution, operations, engineering, and management functions. The automated factory is simply a modular production function unit within a larger computer integrated business functional system.

Prospects for automation of the factory as well as the larger business functional system pose major societal and economic questions. Of immediate concern is the level of short- and long-term structural unemployment that may be associated with workplace automation. Several simplistic and limited evaluations of the problem have already been conducted. A more comprehensive approach to the problem is required and should include considerations of the technology transfer cycle and its relation to technologically induced changes within the economic and social environment.

ANNALS, *AAPSS*, **470**, November 1983

Design of the Automated Factory: More Than Robots

By GERALD NADLER and GORDON H. ROBINSON

ABSTRACT: This article emphasizes the dynamic and complex nature of a robotic factory. Every step from conception to eventual destruction is for, by, and because of people. Twelve principles of design are discussed, with emphasis on the uniqueness of each design and the processes and procedures involved in continual redesign. The technical system is related to the purpose, product, and properties of the social system; there is no optimal technology in itself. The article goes on to examine the requirements of the social system in future automated factories. Experiences in current automated factories, the organizational functions required to support the technical system, and the turbulent environment in which organizations are increasingly forced to operate are reviewed. The kinds of people needed will be required to exercise responsibility and discretion to replace supervision and prescription. The article discusses the usual myths about change, which the design process seeks to overcome, and ends with a note on the appropriate education for future designers, observing that our schools are probably not producing many appropriately trained people.

Gerald Nadler is a professor of industrial engineering and a faculty associate member of urban and regional planning at the University of Wisconsin—Madison, and he serves on the board of directors, Intertherm Inc. He has written over 150 articles and nine books; his latest, The Planning and Design Approach *(1981), was named 1983 Book of the Year by the Institute of Industrial Engineers. He has held five invited visiting professorships, and is on the board of editors,* Design Studies. *Purdue University gave him a Distinguished Alumnus Award in 1975.*

Gordon H. Robinson is a professor of industrial engineering at the University of Wisconsin—Madison. He initiated and heads a pioneering human systems graduate program concerned with sociotechnical systems and quality of working life. He has published in the areas of human factors, human performances, accidents, and sociotechnical systems. Nationally known for his work in forensic human factors in the products liability field, he has received teaching awards, and one of his papers has been selected as Paper of the Year by a national professional organization.

A major difficulty with the topic of future factories is that the mind usually grasps it visually, as a static picture, usually of a specific time—the year 2000 has been popular. This conceptual limitation is significant in debates and planning regarding such new technological possibilities as robots—machines to reproduce human motions and activities.

This snapshot view of the future factory is at best incomplete. It ignores continuing developments in technology, such as those that produced the robot, and it encourages debate about the desirability of specific renderings of technological possibilities, forms unlikely to appear in any event, far less to be influenced by the debate.

We focus here on the process of the design of future factories, a topic far more significant than any specific technological possibility, such as the robot, or for that matter any specific picture, such as the totally automated factory. Three issues must be considered:

1. A factory does not appear suddenly in full operational maturity, but rather is continually designed and redesigned, implemented, constructed, and rebuilt.

2. It is not automatically programmed to improve. Continual energy and direction must be employed if it is to adapt successfully to changing needs and potential.

3. It will never be entirely free of people, never be completely automatic or robotic in this sense.

The factory is a human phenomenon. Every step from conception to eventual destruction is for, by, and because of people.

DESIGN PRINCIPLES

There are several principles to be followed in the original or modernized design of a factory:

1. Each factory, future or present, is unique. Design should not proceed by trying to install a solution from another factory. Even though it may appear, for example, that all hospital technology is identical, there are so many variations in product, resources, and human involvement that one cannot assume that even the standard activities—open heart surgery, food preparation, nursing care patterns—can really transfer easily from one hospital to another. Similarly, simply because General Electric has automated much of its own manufacturing as a showcase does not necessarily mean that it would succeed in promoting sales of factory automation equipment.[1]

2. The hierarchy of the factory's purposes should govern all parts of original or modernized design. Current activities are an insufficient basis for developing factories. Merely automating what is now being done can mean doing very efficiently what ought not to be done at all!

3. The guide for developing a recommended factory design should be based on what the factory would be if it were possible to start afresh, not on what presently exists. Once the purposes of the factory are identified, one should look toward the most effective solution as the target to govern today's design

1. L. Landro, "GE Promotes Factory Automation, but Some Doubt Big Market Exists," *Wall Street Journal,* 21 Oct. 1982, p. 35.

decisions. This guide is often called the "plant after next" concept.[2]

4. Many alternatives—as ideal as possible—should be developed and kept as viable options for as long as possible.[3] As enticing as a robotic factory may appear, a focus on that solution alone is likely to overlook many other more beneficial ways of combining machines and humans.

5. Specifying a plant after next or feasible ideal solution target should be based on regularity conditions rather than all possible conditions. Regularity conditions are usually those external factors that occur most frequently—for example, size of population served, most ordered model—or which are considered most important—for example, most delicate model. Such a concept for a target serves as a guide for later continual changes.

6. As irregularities are incorporated into the target guide, as much as possible of the target should remain in the solution. In other words, do not throw away what is good for 75 percent of the needs simply because the other 25 percent is not accommodated. It may be that more than one factory method is necessary to provide an optimal balance of factors; see also item 9.

7. Each factory, and subunit, should be considered a system. Each system is specified by eight elements, each one defined by six dimensions. This element/dimension idea of a system is illustrated in Table 1. In addition each element in the factory system is itself a system. For example, a robot is considered a physical catalyst in the factory system, but it can likewise be specified in terms of elements and dimensions: the robot's purposes, inputs, outputs, and so on.

8. Information gathering during design is based on these considerations—purposes, ideal systems, regularities—not by the vacuum cleaner approach of collecting all data. Designing a future factory by thoroughly analyzing everything, in order to find out what is wrong with the present system, provides only an inward-looking guide.

9. The specifications of a factory should accommodate all stakeholders—users, customers, workers, and so on. This means that a factory solution will incorporate pluralistic and multichanneled flows or subsystems, each at the appropriate technology level. Thus 25 percent of the production may be robotic, 35 percent semiautomatic, 25 percent job-shop methods, and 15 percent custom or hand work. In other words, there is no one best way. In addition factories will need frequent modifications as products, processes, and so on change. What was once robotic produced may need to be changed as sales change or other technology becomes available.

10. People in a factory should be continually involved in design and modernization. To assume that the automated factory is the last word in design does not recognize the need for continual productivity improvements. Thus productivity circles, workshop groups, and related participatory activities are essential even in the robotic factory.

2. R. Malpas, "Engineering Excellence in Manufacturing Processes," *Engineering* (England), June 1978, pp. 563-68.

3. E. R. Alexander, "The Design of Alternatives in Organizational Contexts: A Pilot Study," *Administrative Science Quarterly*, 24(3):382-404 (Sept. 1979).

TABLE 1
SOLUTION FRAMEWORK–SYSTEM MATRIX*

Elements	Dimensions
Purpose: mission, aim, need, primary concern, focus	Fundamental: basic or physical characteristics—what, how, where, or who
Inputs: people, things, information to start the sequence	Values: motivating beliefs, global desires, ethics, moral matters
Outputs: desired—achieves purpose—and undesired outcomes from sequence	Measures: objectives—criteria, merit and worth factors—and goals—how much, when, rates, performance specifications
Sequence: steps for processing inputs, flow, layout, unit operations	Control: how to evaluate and modify element or system as it operates
Environment: physical and attitudinal, organization, setting, and so on	Interface: relation of all dimensions to other systems or elements
Human agents:† skills, personnel, responsibilities, rewards, and so on	Future: planned changes and research needs for all dimensions
Physical catalysts:† equipment, facilities, and so on	
Information aids:† books, instructions, and so on	

SOURCE: Adapted from Gerald Nadler, *The Planning and Design Approach* (New York: John Wiley and Sons, 1981), p. 89, fig. 9.3.

*Each system is specified by eight elements, each one defined by six dimensions.

†Agents, catalysts, and aids help process inputs into outputs without becoming part of outputs.

11. The specifications of a factory should include only the minimum number of critical details and controls, regardless of how many robots and automated devices may be present. People who operate in the factory system should be given flexibility in achieving their assigned purposes.

12. A solution should be defined as a change plus a delta change. A factory of the future is not itself the solution. It represents merely the change made now, within which a schedule for improvement—the delta change—is also set up.

These 12 principles are in effect guides to the design of any kind of factory, not just an automated one. They provide operational tools for putting into action the three principles with which we began.

THE FACTORY AS A TECHNICAL SYSTEM

The technical production or service delivery system lies at the heart of the organization. It is the breadwinner. No organization can survive if this system fails to produce something of adequate value. Adequate value is not a simple issue, however, as organizations operate increasingly in complex, interconnected environments that are changing rapidly and unpredictably. Many technical systems envisioned just prior to the energy crises of the 1970s now look quite obsolete. Long-range survival now demands technical decisions and plans much more complex than those of the time when most of the currently operating factories were built.

There are four sets of decisions to be made:

—the need or purpose to be achieved by the organization;

—the product or service to be offered;

—the technology available for the production of this product; and

—the specific technical production system chosen.

The first three are highly interconnected. Products may have to await technological possibilities; new developments and products depend on the needs the organization seeks to fulfill; and new technologies may suggest new products. The fourth, how we will render the technology in our specific organization, presents many more choices than commonly realized. The choice here depends on the characteristics of the social system required to operate the technical system and on a wide range of economic factors. It is only now being learned by engineers and managers that specific technical systems are not uniquely specified by the product and technology. There is no one best way at this decision level.

Therefore the purpose of the technical system is critical: our business is to produce writing instruments, not make pencils; to produce starters, not manufacture buggy whips; to distribute products to dealers, not load trucks.

The technical-system part of design identifies state changes in the product or functions at each unit operation. The functions of the product, rather than the activities of people or machines, are what guide design. Thus a factory system represents, at the technical level, a series of unit operations or basic conversions in the materials that are to become part of the factory output.

As a result there are some reasonable guidelines for viewing the factory as a technical system:

1. Products should achieve purposes.

2. Product design is a technical manifestation of human and social needs and required technology.

3. The materials needed are those on which conversion is to take place.

4. The purpose of a manufacturing process is to utilize the materials in item 3 with the most effective benefit-cost methods to arrive at the product desired (item 2). This means that the factory is not necessarily going to be automatic!

5. Control of possible variances in the factory system should be exercised at the earliest unit operation. This avoids later problems. Quality and cost controls are thus often put in the hands of people who work in early unit operations rather than those who produce after-the-fact quality and cost reports.

6. The physical equipment—including such process control devices as computer aided manufacturing—and arrangement—layout, and so on—is the physical/technical manifestation of the process and variance control (items 4 and 5).

7. The information flow system, including computers, that integrates and controls the technical operation and monitors human/social performances is not necessarily going to be automatic:

With the aid of computer controls, machines can now switch from one pre-set tool-and-die configuration to another without trips to the tool room or trial runs. But making it work for you is not [straightforward]. It requires both capital and imagination . . . a thorough rethinking of plant flows, layout and line-balancing logic. The role of workers is also important. Greater flexibility and intelligence are demanded of both people and machines. . . . experimentation is necessary: radical systems that work take time to develop.[4]

P. G. Gyllenhammar[5] addresses these issues, and his Kalmar Volvo factory illustrates some of the choices to be made. At Kalmar two major technical decisions were affected by considerations of the present and future social systems that would operate the plant. The architectural design represented a specific technical design chosen from the technological possibilities. The method of conveying the automobiles being assembled was not among the common technical system choices and required research leading to a new technology of computer controlled carts replacing the assembly line. It is important to note that neither of these decisions was related directly to the product, nor for that matter to the usual economic measures. Both were in fact rather costly and involved some risk. The appropriateness of these specific decisions is not the issue here, of course. The questions asked and issues raised concerning the social system and long-range survival are the issues of interest.

Gyllenhammar feels strongly that large, multinational organizations have two areas of responsibility if they are concerned with long-range growth, or even survival. One is responsibility for the people, their social system. The quality of life for these people, now and in their future, affects Volvo and is affected by Volvo through what is now commonly known as quality of working life. The other responsibility is toward the material resources that society chooses to make available for, in Gyllenhammar's case, the production of automobiles.

It has thus become more clear that what we used to think of as decisions based purely on our proprietary choice of product and the correct, optimal technology are not that kind of decision at all.

THE FACTORY AS A SOCIAL SYSTEM

The automated factory will depend more, rather than less, on the people within it. While there will be fewer engaged in direct production, there will probably be more people in allied, supportive functions. That is, the production function will be more richly interconnected with other functions and organizations, and more dependent on them. Perhaps the more interesting issue, however, is what these people will be doing, and consequently what skills and organization they will require.

Twenty years ago, in a well-reasoned article, N. Jordan[6] spoke of the need for people in automated systems to provide manual backup in the event of machine failure. The importance of this function became dramatically apparent within a few years as several U.S. space missions survived only because of interventions by their on-board humans, who up to then had been viewed by many designers as simply cargo. As we have moved

4. T. M. Hout and G. Stalk, Jr., "The Big Revolution on the Factory Floor," *Wall Street Journal,* 12 July 1982.

5. *People at Work* (Reading, MA: Addison-Wesley, 1977).

6. "Allocation of Functions between Man and Machines in Automated Systems," *Journal of Applied Psychology,* 47:161-65 (1963).

farther along in the sophistication of our technical systems we have also moved farther from any significant possibility of manual backup operation. This is particularly true in manufacturing, where the nature of the product, the resources for its production, and particularly the necessary rates of production prevent almost all possibility of human intervention in the production flow.

What then are people doing on the shop floor as we move into highly automated factories? We can begin to examine this question by looking at our most advanced and technologically sophisticated factories. One of these is a chemical processing plant discussed by L. E. Davis and C. S. Sullivan.[7] In this plant 150 employees are involved with $300 million in capital investment, $2 million per worker. This ratio in itself begins to suggest some issues for the design of the social system. What responsibility does this worker have for his or her $2 million machine? Or in terms of costs, what can you afford to pay for, or not afford not to pay for in a social system that is so relatively inexpensive?

The design of this plant, and a number of others with similarly high degrees of automation—such as food processing and paper making—was greatly influenced by an emerging body of conceptual knowledge and technique known as sociotechnical systems.[8] Starting with studies of the impact of new, highly mechanized machines in the British coal mines in the late 1940s, extending these insights to include new roles for supervisors in experiments with automated weaving looms in India in the early 1950s, and continuing through important new ways of characterizing the interactions between the technical and social systems with studies in paper making in Norway in the mid-1960s,[9] this theory has turned aside much conventional wisdom in regard to both individuals and the organizations in which they are embedded. Of greater importance to the issues before us here, however, is that the sociotechnical view gives us increasing understanding of what the social part of the system is in fact there for. Prior to this new understanding, most designers simply viewed the people as doing—temporarily—that which they could not yet automate. Forty years of experience with automation have highlighted people's curious tendency to stick around. We are now beginning to understand why.

A. B. Cherns and G. J. Wacker[10] discuss the potential extension of Parsons's macrosocial model to the industrial organization. They present four functions as necessary for the continued survival of the organization: (1) production—they use the term "goal attainment," but we have found it unnecessarily confused with goal setting in discussions with industrial managers; (2) maintenance and development; (3) adaptation; and (4) integration.

Production is the breadwinner. It is this function that adds value to some resource that in turn can be sold at a profit. The new roles for the social system in highly automated production are

7. "A Labour-Management Contract and Quality of Working Life," *Journal of Occupational Behaviour,* 1980, pp. 29-41.

8. E. L. Trist, *The Evolution of Socio-Technical Systems* (Ontario: Ministry of Labour, 1981).

9. P. H. Engelstad, "Sociotechnical Approach to Problems of Process Control," in *Design of Jobs,* 2nd ed., ed. L. E. Davis and J. C. Taylor, (Santa Monica, CA: Goodyear, 1979), chap. 19.

10. "Analyzing Social Systems," *Human Relations,* 31 (1978).

now fairly well understood. These roles center on variance control—setting right that which has gone wrong, and learning to do it better. The need here for people will be discussed in more detail in a later paragraph.

Adaptation actually encompasses two functions: protection of the production system from outside influences that would reduce its effectiveness and thus the profits; and the introduction of new technology and continued technical system improvements that will enhance the production potential. Adaptation thus is forced continually to balance the need for change with the need to let production alone so that it can produce with high efficiency. Some, such as R. H. Hayes,[11] have noted that control of this balance is possibly one of the things the Japanese do differently, if not better, than we. Decision making within this function obviously incorporates many nonquantitative variables and values. Automation will probably never flourish here, although computers will help greatly with structuring the decisions and projecting the possible consequences.

Maintenance and development provide the organization with an appropriate social system; the major responsibilities here involve new people, their adjustment and training, and management of their careers. The motivations of people at work, their expectations of other people at work and of the organization, and their individual differences would appear to make any systematic automation rather difficult, if not actually incongruent with the maintenance and development function. As with adaptation, however, people performing this function can certainly be aided by computer technology. Of course this function would be unnecessary if the factory had no people—no social system—to maintain.

Integration is of course much of what traditional management does. It is important to point out, however, that newer organization designs, particularly in highly automated plants, have an increasing amount of integration and coordination at lower levels, levels previously thought not properly to contain these functions. This movement of decision making and authority downward, and the often accompanying flatter organizations, results less from considerations of quality of working life, or employee satisfaction, than from the requirements of the technical system, although it usually improves both. It is more difficult to envisage the extent and form of automation here. The computer as an aid to decision making and planning is already with us. Its value will grow. Automating the managerial function is another question and again has a degree of incongruency if any people remain to be managed. Goal setting of course must remain a human function.

Returning to the production system itself, the notion of the social system here controlling variances has already been introduced. This concept has been very helpful in understanding the form and needs of the social system in highly automated plants. It is important to point out that we cannot, in any but the most trivial of production systems, reach that point of eliminating or automatically controlling all of the variances. Continued evolution of product and technology alone preclude this possibility. In the chemical processing plant

11. "Why Japanese Factories Work," *Harvard Business Review,* Jul.-Aug. 1981.

previously mentioned[12] an examination of this issue led to the design of a learning organization, such that control of things gone wrong or new demands would be as effective as possible. It is worth mentioning at this point that the highly automated technical system originally planned for this plant was deliberately degraded in order to facilitate this learning. This singular act may foretell a very different future for automation than technological possibilities alone would indicate.

Organizations are increasingly forced to function in what E. L. Trist[13] has described as a "turbulent" environment. He suggests that the appropriate design for success, or simply survival, in this environment is one increasingly looking toward rapid adaptation. Adaptation to change is done fairly well by people and rather awkwardly, at best, by technical systems. Automation of response to changes, which strike all of the systems' functions as previously described, will remain difficult—actually not even appropriate in the sense that the factory is to serve the people, and of course the people's needs and desires are part of the continuous environmental change.

DEMANDS ON THE PEOPLE IN THE FACTORY

What the people are to do has already been outlined. The next question for the designer of the automated factory is whom these people should be. It is already clear that scientific management's replaceable-parts person will not do even in today's highly automated factories. It is perhaps fortunate then that such people are in short supply today.[14]

The needs here become clear as we see that the role is replacing the job.[15] Role includes the social system's rich communications connections, which become increasingly necessary with automated technology. With this role we also see an increasing need for responsibility. No longer can the operator wait for supervisory request. The discretionary parts of the job increase, while the prescribed parts are reduced. Even today we systematically underestimate the importance of worker discretion, a fact painfully recognized by the British Rail System as their operators brought it to a halt with "work to rule."

Within the general concepts of role and responsibility the functional needs for the social system point to the more specific characteristics that the people must possess. The control of the variances in the technical system requires three things in addition to responsibility: skill, knowledge, and the willingness to exert authority that has been delegated.[16] Knowledge, as opposed to a collection of skills, seems increasingly necessary with advanced technology—knowledge not such as that related to school but rather, or in addition, knowledge of the product or service being produced and of the quantities, qualities, and costs necessary for the organization's continuation or growth.

The adaptation social system can be conceptualized as the people controlling

12. Davis and Sullivan, "Labour-Management Contract."

13. "The Environment and System-Response Capability," *Futures*, Apr. 1980, pp. 113-27.

14. L. E. Davis, "Individuals and the Organization," *California Management Review*, 22 (1980).

15. Davis and Taylor, eds., *Design of Jobs*.

16. L. E. Davis and J. G. Wacker, "Job Design," in *Handbook of Industrial Engineering*, ed. G. Salvendy (New York: McGraw-Hill, 1982), chap. 2.5.

the variances in production and certain staff functions, industrial engineering being prominent. The new needs for this engineer will be discussed later. The integration social system now extends from the bottom to the top of the organization. Cooperation through the understanding of common goals and assumptions becomes critical as these flatter organizations move to a mode of "us against the product" rather than "us against them." The development social system must of course provide and maintain all of the preceding and therefore must have a keen understanding of all of these new social system needs, as well as the skill and knowledge to produce the necessary people.

HOW TO PROCEED WITH DESIGN

This section will seek to relate and integrate this volume's concern—the futuristic robotic factory—with the considerations of how to design a factory, as presented in this article. We believe it is far from sufficient to adopt the snapshot image of the factory of the future, because it portrays a time so far distant from now that it does not consider the day-by-day, week-by-week, month-by-month efforts needed to move from a need or purpose to a specific facility for producing the outputs. Each robot will have to be evaluated on its own merits; each set of robots will have to be evaluated in the same manner. The normal evaluation processes for any capital expenditure will tend to reduce greatly the number of places where a robot might be used, simply through the normal dictates of risk taking and return on investment.

We would go so far as to claim that most of the projects of the robotic factory will no more come to pass than the images of the 1940s concerning what our cities would look like came to pass in the 1980s. There has been a great deal of literature that criticizes the idea of global models, world futures, and similar projections as being expressions of visionary thinking rather than implementable change. On the other hand, there is every reason to maintain the vision of the robotic or automatic factory. As the quintessential plant after next it certainly operates as an excellent guideline, and stimulates thinking about the design process.

In essence the process for design must consider, for the overall process as well as for each of the individual unit operations and the connections, the following iterative steps.

1. Determine the purposes to be achieved within the larger hierarchy of possible purposes. Focus on missions and aims rather than equipment and activities.

2. Develop a "philosophy statement" of the organization's values and theories.[17]

3. Develop ideal systems for achieving the purposes within the philosophy statement. Here is where the dreaming of the future states can be incorporated. In addition, however, other kinds of arrangements can also be considered, even ideal arrangements of humans and machines.

4. Develop a feasible ideal system target for the regularity conditions. Work out the plant after next based on the most important considerations rather than simply trying to incorporate all models or sizes or colors at the same time.

17. A. B. Cherns, "Can Behavioral Science Help Design Organizations?" *Organizational Dynamics,* Spring 1977.

5. Develop the details for the recommended solution, including the irregularities. This may result in more than one production or information flow pattern identified in the target solution. Thus it is likely that the regularity conditions may end up having a robotic or automatic segment or factory, whereas the irregularities may have semiautomatic or other types of systems to accompany the robotic line.

6. Install the factory, which entails the whole range of obtaining approval, setting up construction, purchasing, training, measuring expected performances, monitoring, and evaluation.

An important ingredient of this planning and design approach[18] is that four other factors must be considered simultaneously with these six steps; all four must be integrated in the considerations of each of the steps, rather than being treated independently.

1. Specifying and presenting the solution. This involves utilizing the systems view of Table 1 as a definitional framework to make sure the details are thoroughly considered.

2. Involving people. Experts in engineering or facility planning should not be the sole determiners of the factory. A wide range of stakeholders are almost always necessary. One factory being developed as a joint venture by two very large firms hired up to 20 percent of the anticipated factory personnel over a year before the factory was built and constructed. The main purpose was to get them involved immediately in the design of the factory in which they were to work.

3. Using information and knowledge effectively. Most research results are too narrow and lack a system perspective. Frequentist statistics, the current basis for reporting significance of research efforts, are not useful in predicting the specific instance or performance needed in the design process. Subjective probabilities—Bayesian statistics—for assessing likelihoods are a better framework, for example, for how to use information and knowledge.

4. Arranging for continuing change and improvement. For example, policies of the organization regarding employment security, continuing education of people, involvement of individuals in productivity or quality circles-type activities, and system research and development are crucial factors to consider as one goes through the design steps. Thus policies that do not look far enough ahead—for example, no policy on job security when work is to be eliminated—can negate the benefits of almost every step in the design process. Good policies, on the other hand, can enhance and motivate people's involvement in each of the steps. These types of policies are considered part of what is called quality of working life, and they help to incorporate the social system aspects of a factory as described before.

What this design process seeks to overcome are the usual myths that surround the development and implementation of significant changes.

Myth 1: poor attitudes cause poor implementation of the robotic factory. Most notable among those charged with this attitude are the labor unions. Yet the poor attitudes most often are symptoms and measures, not the substance or the underlying causes. Most often the poor attitudes reflect ineffective management practices. The design process seeks to embed all such factors within

18. G. Nadler, *The Planning and Design Approach* (New York: John Wiley and Sons, 1981).

the overall consideration of a total system.

Myth 2: high technology and capital expenditures provide the only route to dramatic increases in productivity. As valuable as high technology and capital expenditures will be, each situation needs its own solution—its own appropriate technology—not just the solutions developed by our high-productivity competitors.

Myth 3: adopt the robotic factory program that works elsewhere. The characteristics of a given situation—type of product, sales volume, and so on—even if they are nearly identical to those of another situation, do not extend to such important issues as the people as individuals, the history of the organization, and the style of management. These are always different.

Myth 4: developing automatic factories is so specialized that we should entrust it to the experts. So many analysis techniques and production devices are available that experts are needed to determine which ones should be used and what data should be collected. But this ignores how completely intertwined a factory is with every problem: it is an aggregate concept rather than just a technical one.

Myth 5: set up a separate automatic factory department. This is bureaucratization of the expert myth. A department most often brings people together where they are more likely to interact only with each other, thus becoming isolated from the rest of the organization.

Myth 6: decree the establishment of an automatic factory and it will happen. Managements that make this decision often assume that further planning is unnecessary, since now implementation is simple and self-evident. Again arrival at significant change needs day-by-day, week-by-week, and month-by-month planning with the design process already discussed.

IMPLICATIONS FOR THE EDUCATION OF INDUSTRIAL ENGINEERS

This article has argued that as technology has become more complex and automated, the social system required to operate and support it also becomes more complex. Roles replace jobs; coordination, cooperation, and knowledge demands affect all of the people. We therefore have more complex technical and social systems, and they are more richly interconnected. Not only is the future factory, therefore, a more complex system, in the systems theoretic sense, but it is increasingly viewed as an organic rather than a deterministic structure—more like a forest than a piece of clockwork. The environment is all pervasive, and the factory itself is continually learning and evolving. For example, Fanuc—the factory in Japan claimed by many to be the most advanced illustration of robotic manufacturing—has signs all over the plant exhorting the 60 to 70 employees to cut total costs by 30 percent!

The design, and the management, of this system requires skills, tools, and strategies that are not in great supply—and for that matter not even well defined. The responsibility thus rests on our educational system to uncover these needs and produce the needed designers and managers. The design issue we are addressing here, of course, is the total system design, recognizing that the system is more than its parts; we are in fairly good shape, educationally at least, for the parts design.

We hesitate simply to say that we need systems designers, for according to job descriptions or degree program credentials, these people abound. They come from computer science, electrical engineering, and industrial engineering departments. Few of these people today have the credentials we actually find necessary. The numerous failures of orderly and timely implementation of new technological ideas or new organizational forms is the clue. To most current engineers, the term "system" means a fairly complex machine, a mathematical algorithm, or a computer.

Four issues are relevant.

1. The system is more than its parts, and cannot be understood, much less designed well, if this is not explicitly taken into account.

2. The system is sociotechnical with interacting technical and social systems, and is virtually guaranteed to be suboptimal if either part is considered in isolation.

3. The design is never finished, partly because of the continuing advances in technology such as automated machines, and partly because change takes time, energy, and resources.

4. Implementation of the design requires a design of its own.

These are the issues the designer must understand and have mastered the necessary tools and techniques to effect. Such a designer is then a generalist, for it would be quite impractical to master all of these issues and be a good electrical engineer or computer scientist. This designer will coordinate, facilitate, organize, and plan. Subsystem designers serve in specific phases of the design process as their own specialty is needed. Probably the best current analogy is the architect, with the building as the total system. Even the architect is a subsystem designer for our new factory, however, as is nicely illustrated by Volvo's Kalmar building design.[19]

This designer will also coordinate the needs for the total range of stakeholders, a set much larger than in the past. Government agencies, community groups, consumer agencies, unions, and so forth will have a stake in the design, and as such will have the ability to offer assistance, or bring our plans to a complete halt—this latter result having been seen often in recent years in regard to freeways, dams, and power plants.

This professional systems generalist is not being produced in many places today. Pieces of the appropriate course of study are in effect in some engineering and architecture schools. There is fair agreement that practice must be woven into the curriculum, or at least case studies of greater depth and longitude than most. Perhaps lessons can be learned from those producing generalists in medicine or law. What is clear, however, is that the lack of these people will substantially impede our progress toward whatever the factory of the future will be.

19. Gyllenhammer, *People at Work*.

ANNALS, *AAPSS*, **470**, November 1983

Impacts of the High Technologies on Vocational and Technical Education

By WARREN H. GROFF

ABSTRACT: The industrial nations of the world are in the turbulent times of a structural shift from an industrial society to a technological society. This shift is evident in the unprecedented rate and scope of change in science and technology. The infusion of high technologies that constitute the automated factory or the factory of the future is having and will continue to have a profound impact on vocational and technical education. Institutions of secondary and postsecondary occupational education must develop processes to (1) plan and manage strategically, (2) upgrade curriculum and methods of instruction to reflect state-of-the-art technology, (3) develop programs that focus on human resource development, and (4) demonstrate an adequate return on investment in terms of value added to individuals and society. These issues must be addressed if there is to be a closer relationship between occupational education and the economy, and if this nation's people are to be the beneficiaries of science and technology.

Warren H. Groff is vice-president for academic affairs, North Central Technical College. He chairs a task force on high technology for the chancellor of the Ohio Board of Regents and is a member of the Higher Education Telecommunications Committee. Dr. Groff served on the National Conference Committee of CAUSE as a program track coordinator in 1979, 1980, and 1981; was the 1970-80 president of the Technical Education Division of the Ohio Vocational Association and received its 1981 Distinguished Service Award; and served on the American Vocational Association Technical Education National Conference Committee, 1982 and 1983. The recipient of many awards, he was one of two faculty at the Snowmass Institute on Strategic Planning and Management, 1981-83, and assisted in conducting an American Council on Education Leadership seminar for the newly created Massachusetts Board of Regents, 1981.

THE automated factory or the factory of the future will have profound impact on vocational and technical education. In the next decade no issue is more important than the relationship between occupational education and the economy. In the past occupational education has seen its relationship to the economy primarily in terms of providing a preentry work force by a group processing delivery system. That emphasis was broadened to include some continuing education, delivered usually in the same way. Recently new forces have appeared. These forces include foreign competition; rapid technological advances; changes in productivity; high costs; human, plant, and entire industry obsolescence; and infrastructure deterioration. These forces have created massive dislocations in our economy and in the economy of each industrialized nation in the Western world. The focus on a preentry work force and continuing education will continue to be important—but not sufficient—in the future. New expanded relationships will be required between occupational education and the economy in the computer-literate, high-technology, information society.

Before discussing these relationships, however, it is important to clarify the world of occupational education. The occupation structure of the United States can be described on a continuum ranging from unskilled workers on one end to professionals and high-level managers on the other end. Professional education is theory oriented and focuses on producing researchers or theoreticians and managers or professionals. Between the unskilled workers and the professional education ends of the continuum are vocational education and technical education. Vocational education is practice oriented and focuses on the preparation of craftsmen and skilled workers. Technical education blends theory with practice to prepare a broad range of paraprofessionals, specialists, technicians, and middle managers. Educationally, vocational education tends to be regarded as secondary education, and technical education usually refers to postsecondary education but not to the level of the baccalaureate degree.

Impacts of the automated factory on vocational and technical education will be discussed in terms of (1) strategic planning and management, (2) curriculum and instruction, (3) human resource development, and (4) accountability and leadership.

STRATEGIC PLANNING AND MANAGEMENT

During the post-World War II years institutional planning focused on acquiring more resources and building facilities for the increased number of students resulting from the equal-right demand for access to postsecondary education. Planning in postsecondary education during the 1960s responded to immediate needs with minimum regard for the future. During the 1970s the influx of traditional students, 18 to 22 years old, began to stabilize. Many private and public senior institutions experienced the impact of a broad range of demographic, social, political, and economic forces. As a result organizations such as the Council of Independent Colleges—formerly the Council for the Advancement of Small Colleges—the Academy for Educational Development, and the American Association of State Colleges and Universities launched programs relating to comprehensive institutional planning. These projects and others like them all stressed the need to assess the external environment. The literature began to reflect descriptions of institutional planning processes, including some way to assess the external environment.[1]

During this period phenomenal growth occurred for a broad range of education and training providers, including business and industry, the Department of Defense, professional associations, adult education associations, and proprietary organizations. The National Conference Board indicated that in the single recession year of 1975 this nation's 7500 largest private employers spent over $2 billion on employee education, or as much as the recent annual totals of all contributions from all sources to colleges and universities.[2] In 1979 an article in the *New York Times* stated, "The American Telephone and Telegraph Company spent $700 million on educational programs for its employees, or more than three times the $213 million annual budget of the Massachusetts Institute of Technology.[3] An article in the May 1980 issue of the *Training and Development Journal* stated, "Industry spends on employee education more than six times the amount appropriated by all the states for all of higher education."[4] In 1981 an article in the *New York Times* stated, "Within a short drive of Boston, a city with no shortage of higher education, are four new degree-granting programs that are not even affiliated with a college or university. They are sponsored by a hospital, a bank, a consulting firm, and a computer manufacturer."[5]

Thus the stabilization in the number of traditional students and the intrusion of a broader range of education and training providers caused changes in institutional and system planning processes. These processes began to embrace the strategic concept. Strategic planning is essentially a way to match an organization's resources to a changing environment. It requires a way to develop a conceptual framework about the changing nature of society, audit the strengths of the institution, and match institutional strengths with opportunities in the external environment.

Conceptual framework for society

The history of the development of human society can be traced from the hunting society through the agricultural society to the industrial society. In the hunting society mankind was concerned primarily with extracting things from nature. The transformation to the agricultural society was slow and based on rather simple technological innovation. The hunting and agriculture societies can be characterized by interactions between people and nature. In contrast the transformation from the agricultural society to the industrial society occurred more quickly and resulted from technological advances in energy, transporta-

1. Warren H. Groff, "Strategic Planning: A New Role for Management Information Systems" (Paper presented at the 1981 CAUSE National Conference, 2 Dec. 1981).

2. Carolyne Davies and Austin Dougherty, "Our 14,000 Colleagues in Business Industry and Government," *AAHE Bulletin,* Oct. 1978, p. 7.

3. Stan Luxenberg, "Business is Big in Education Too," *New York Times,* 7 Jan. 1979, sec. 13, p. 15.

4. Robert J. Kost, "Competition and Innovation in Continuing Education," *Training and Development Journal,* May 1980, pp. 48-67.

5. "Business Is Cutting into the Market," *New York Times*, 30 Aug. 1981, sec. 12, p. 1.

tion, communications, raw materials, and research and development networks. The industrial society can be characterized by interactions between people and goods or fabricated nature. More recently advances in the industrial society have resulted from the integration of macro technological systems, the aggregation of complex technological developments in each of these networks.

During recent years we have experienced the onset of a transformation to the so-called information society, a shift from physical productivity of material goods to information productivity that can be expected to bring about fundamental changes in human values, in trends of thought, and in the political and economic structures of society. This learning and information society will be characterized by interactions between people, ideas, and knowledge.

Implications for occupational education

The implications for occupational education are apparent. Institutions as well as systems need some way to monitor demographic, social, economic, and governmental planning forces to develop visions and scenarios of possible alternative futures for their institution. Tools for doing this include (1) needs assessment, (2) market analysis, (3) environmental scanning, (4) trend analysis, (5) policy analysis, and (6) issues management. Visions and scenarios should be based on hard data about the institution or system and its service area. The analysis should attempt to match institutional strengths and weaknesses with opportunities in the external environment. Given such monitoring and analysis, an institution/system may develop the most likely scenario of its future and then translate that scenario into specific goals and objectives to which resources can be allocated.

Assessing the external environment

Anticipating unprecedented change in societal trends and values in the 1980s, the Resource Center for Planned Change of the American Association of State Colleges and Universities developed *A Future Creating Paradigm* as a way of planning futures and bring planning assumptions into focus.[6] The project uses a cross-influence matrix of 12 societal trends and 12 values to determine goals in 10 areas. The 12 societal trends are population, government, global affairs, environment, energy, economy, science, technology, human settlements, work life style, women, and participation. The 12 societal values are change, freedom, equality, leisure, foresight, pluralism, localism, responsibility, knowledge, quality, goals, and interdependence. The 10 goals are finance, students, research and development, public service, facilities, faculty, curricula, administration, resources, and athletics.

The central point to this discussion is that strategic planning and management must include a critical analysis of forces and trends that are essentially external to the organization. Data processing and management information systems in the past have tended to focus on data elements relating to internal operations of the institution, such as registration, scheduling, class rosters, space utiliza-

6. *A Futures Creating Paradigm* (Washington, DC: American Association of State Colleges and Universities, 1978).

tion, grade reporting, student aid, payroll, budgeting, and other administrative applications. Although the processing of data elements that are essentially internal to the institution will continue to be an important factor in decision making, organizational strategic planning processes are becoming increasingly more dependent upon data elements external to the institution.

Aggregate categories of data about the external environment include (1) demographic trends, (2) economic trends, (3) social indicators, (4) governmental planning, (5) technological advances, (6) changes in the workplace, (7) energy requirements, and (8) value shifts. Sample subcategories for the first three of these groupings are displayed in Table 1.

CURRICULUM AND INSTRUCTION

The curriculum is comprised of three formats: (1) content, (2) delivery system, and (3) evaluation. The content format consists of subject matter derived from the theoretical base or methodology in a discipline or combination of disciplines. Robotics, for example, combines principles of electronic and mechanical engineering. The delivery system format consists of the different designs used to facilitate student learning. The mosaic of designs range from teacher-centered formats such as the lecture method through learner-centered formats such as teacher-free automated learning systems. Evaluation formats consist of the distinctive way of measuring student learning outcomes in the cognitive, psychomotor, and affective domains. Evaluation formats range from simple low-order cognitive factual information recall tests to demonstrations of competency in a clinical laboratory internship environment, which requires the integration of higher order cognitive, psychomotor, and affective domain competencies. A curriculum is a set of activities that, operating collectively, is expected to achieve a well-defined set of goals and objectives within a specified time frame. The activities in a program should be a set of individual parts that have a carefully designed interrelationship.

Content formats

Space will not permit a detailed discussion of all the ways the automated factory or the factory of the future will affect curriculum formats. A detailed discussion would require analysis of advances in numerous fields, such as new materials—fiber-reinforced composites, foamed metals, alloys, superconductors, new coatings, and adhesives; measuring devices and sensors—high-speed observation of dynamic processes, chemical reactions, subatromic particles; and specific technologies—communications, computers, aerospace, electronics, information, and robotics. One detailed example about manufacturing will indicate some of the automated factory's impact on the content format.

Most manufacturing activities can be characterized by the quantity of identical products produced as low, medium, or high volume. The word "manufacturing" for many people means mass production or high-volume manufacturing. Actually low- and medium-volume manufacturing—batch manufacturing—represents at least three-quarters of all industrial production. In spite of its importance very little automation has been

TABLE 1
SAMPLE SUBCATEGORIES FOR SUGGESTED DATA CATEGORIES

External Environment		
Demographic Characteristics	Economic Trends	Social Indicators
Population size	Textile industry	Population and the family
Age distribution	Auto industry	Health and nutrition
Sex ratio	Electronics industry	Housing and the environment
Marital status	Telecommunications industry	Transportation
Ethnic and cultural characteristics	Health care industry	Public safety
Education levels	Agriculture industry	Education and training
Economic status	Airline industry	Work
Population density	Energy industry	Social security and welfare
Degree of urbanization	Steel industry	Income and productivity
Racial composition	Insurance industry	Social participation
Unemployment	Shipbuilding industry	Culture, leisure, and use of time
Poverty and deprivation	Biotechnology industry	
Illiteracy	Aerospace/space industry	
Existence of basic community services	Defense industry	
Social, political, and economic well-being	Synfuel industry	
	Mining industry	
	Education industry	

applied to batch processing. Major changes, however, are occurring. To understand these changes, it is helpful to look at contrasting models.

The basic unit for batch manufacturing is a machine with an operator. The role of the worker is to operate the machine by manually adjusting buttons, dials, handwheels, and levers; the machine usually has some sort of electric or hydraulic power. The next advance is electronic intelligence in the form of computer control to run the machine accurately, consistently, and repeatedly, through a complicated series of operations. Currently the item moves on a journey from machine to machine to be processed from raw material to part to component to finished product. The journey takes a long time because the item being produced spends 95 percent of the time as inventory standing in line, waiting, or in transit from machine to machine. New technology such as computers and robots offer opportunities for automating the process.

The transition from today's practice to a fully computer integrated manufacturing process will take a considerable amount of time. State-of-the-art technology is fragmented and consists of individual computer aided functions that provide greater speed, efficiency, accuracy, and repeatability for the individual tasks. Data bases are not uniform throughout the series of manufacturing operations. In the factory of the future, computer aided machines will be sequenced together with automatic handling and transport of materials as they make the journey from raw material to finished product.

This scenario translates into curriculum content implications in numerous ways. First, faculty, as well as students at the vocational and technical education levels, need to understand the conceptual framework of the automated factory. There are essentially three phases: (1) design, (2) planning for manufacturing, and (3) manufacturing. The entire design process will be modified by computer aided design (CAD) and its two major subfunctions, drafting and analysis. Everything beyond CAD can be termed computer aided manufacturing (CAM). Planning for manufacturing includes all the aspects illustrated in Tables 2A and 2B.

Second, beyond a conceptual framework of the automated factory is a conceptual framework of the automated management systems necessary to support the establishment, including the office of tomorrow or the paperless office. All establishments are evolving relative to these conceptual frameworks in an ad hoc or planned manner, recognizing that variations in stage of development exist by type of industry, geographic area, and size. For example, the aerospace, defense, and space industries have benefited from the hugh research and development (R&D) investments by this nation's federal government. Some of the technological advances in these industries are finding their way into other industries. Curriculum content must translate the stage of evolution of establishments in an institution's service area into specific cognitive, psychomotor, and affective domain competencies. Needless to say, computer literacy is rapidly becoming an essential competency in all contexts.

Delivery system formats

Several persons have commented on the educational system and the changes necessary to deliver quality education and training. Francis Keppel, former U.S. assistant secretary of health, education and welfare, talks about *The Necessary Revolution in American Education:*

> The first revolution in American education was a revolution in *quantity*. Everyone was to be provided the chance for an education of some sort. The revolution is almost won in the schools, and is on its way in higher education. The second revolution is *equality* in opportunity. That revolution is under way. The next turn of the wheel must be a revolution in *quality*.[7]

Willard Wirtz indicates that we are in a revolution that is linking the world of work more closely with the world of eduation, a process of integrating the "learning to do" system with the "doing to learn" system.[8]

Leslie uses Toffler's *The Third Wave* to develop an educational institution in a technological society dominated primarily through electronic media:

> This methodology would allow the learner to process at his/her own rate and style, within his/her own time period, at his/her desired location, drawing upon learning materials from throughout the country and the world. Computer science and electronics courses and programs of study would be an integral part of the curriculum. Faculty would be cross-trained in a variety of disciplines and teaching styles. They would have flexible work schedules and loads and might share an assignment with a spouse or colleague. Many

7. Francis Keppel, *The Necessary Revolution in American Education* (New York: Harper & Row Publishers, 1966).

8. Willard Wirtz, *The Boundless Resource* (New York: E. P. Dutton, 1976).

TABLE 2A
PLANNING FOR MANUFACTURING

Capacity Planning	Materials Requirements Planning	Computer Assisted Process Planning	Part Programming	Tool Designs
Systems simulations Equipment requirements Manpower requirements Financial justification	Shop orders Production schedules Capacity requirements	Sequence of operations Time standards Cost estimates	Machine movements Process parameters Cost estimates	Fixture design Mold design Special tools

TABLE 2B
MANUFACTURING

Shop Floor Control	Process Automation
Machine scheduling Machine monitoring Tool scheduling Labor reporting	Process machines Sensors Flexible automation Materials handling Inspection Assembly

SOURCE: Richard Kegg, director, Machine Tool Development, Cincinnati Milacron, 20 Oct. 1982.

faculty would instruct from their home or electronic cottage.[9]

If vocational and technical education is to be responsive to the needs of the learning society, it must accelerate the development of multiple path diagnostic and education and training delivery systems.

Student learning outcomes evaluation formats

Knowles states that the challenge for education is one of a "continued shift away from the knowledge and skill transmission model which we borrowed from traditional schooling toward a competency-development model." The competency-based model developed from emphasis on behavioral objectives and criterion-referenced objectives. That development must continue in collaboration with a closer partnership between the R&D establishment, business and industry, and vocational and technical education. Only through a collaborative relationship can the components of competence be specified for such newly emerging fields as the industrial laser technician, fiber optics technician, industrial robot production or maintenance technician, bionic-electronic technician,

9. Judith W. Leslie, "As the Third Wave Approaches Higher Education: Planning for the Electronic Institution," *CAUSE/EFFECT* 4:1, 6-15 (Jan. 1981).

materials utilization technician, holographic inspection specialist, and hazardous waste management technician.

Integration and articulation

In *Less Time, More Options,* the Carnegie Commission on Higher Education states:

> Society would gain if work and study were mixed throughout a lifetime, thus reducing the sense of sharply compartmentalized roles of isolated student v. workers and of youth v. isolated age. The sense of isolation would be reduced if more students were also workers and if more workers could also be students; if the ages mixed on the job and in the classroom in a more normally structured type of community; if all members of the community valued both study and work and had a better chance to understand the flow of life from youth to age. Society would be more integrated across the lines that now separate students and workers, youth and age.[10]

As the pace of technological change accelerates, no education will last a lifetime. The linear life plan that had been divided into three full-time phases—education for the young, work for the middle-aged, and enforced leisure for the elderly—is yielding to a blended life plan that permits learning, work, and leisure to go on concurrently. Organizational planning processes and personnel policies must incorporate such flexibility for students and high school and college personnel.

Articulation must occur within a program and between programs and levels. Many curricula were organized or packaged for traditional students—those 16 to 18 or 18 to 22 years old who could attend full-time during the day. During the past several years the student profile has changed, but too often curriculum organization has not changed. There are multiple approaches to organizing the curriculum. Two approaches can be diagrammed, say, in the form of two alternative ways of organizing a cafeteria. One approach is the straight line cafeteria with all food items arrayed in a linear, sequential manner. Another approach is the flexible access cafeteria with food items clustered by type. In the first, all persons are required to move past all food items even if the item is not needed or wanted. In the second approach, persons can go directly to items they want.

Research indicates that self-directed learners are distinguished by such characteristics as (1) openness to learning opportunities, (2) self-concept as a problem solver and effective learner, (3) a dedication to initiative and independence in learning, (4) love of learning, (5) acceptance of responsibility, (6) future orientation, (7) creativity, and (8) ability to use problem-solving and basic study skills. If curriculum formats are prescriptive, teacher centered, and arrayed in a straight line cafeteria organization, self-directed learners will derive little satisfaction from their college experience. The problems are even more complex for the non-self-directed learners.

It is not possible to enter into a detailed discussion about articulation between program levels such as (1) secondary vocational education and one- or two-year postsecondary technical education and (2) two-year technical education and other postsecondary education programs. Suffice it to say that this nation cannot afford the current duplication of effort, be it explained in terms of human learning inefficiencies

10. The Carnegie Commission on Higher Education, *Less Time, More Options* (Highstown, NJ: McGraw-Hill, 1972), p. 2.

or economic investments. A key to the reduction or elimination of the problem is the specificity of components of competence that should be the outcome of the learning experience.

HUMAN RESOURCE DEVELOPMENT

Professional preparation and professional continuing education is an extraordinarily complex task today. Most faculty were graduated from undergraduate and graduate programs that focused on service delivery rather than educator roles. Their programs tended to concentrate on the competent performance of some role other than that of learning facilitator. Their programs dealt minimally, if at all, with curriculum content formats, packaging formats, or learning outcomes evaluation formats. Nor did they concentrate on stages of adult development and distinctions between pedagogical and androgogical principles. Just as it is important for elementary school teachers to understand principles of human growth and development for the relatively homogeneous populations they serve, so too is it important for the professional educators in postsecondary education to understand the increasingly heterogeneous populations they serve.

Vocational and technical education personnel pass through various stages of career development. This is true for faculty, staff, and administration. Just as organizations need a diagnostic/developmental system to grow to fully maturity, so too do individuals need a diagnostic/developmental system to help them understand themselves in the formation and revision of personal and professional goals and ways of achieving them. If lifelong learning is to progress from an idea to a full-blown reality, it would appear that occupational educators should lead the way. It would also appear incontrovertible that maximum synergism is achieved when individual diagnostic/developmental systems are in harmony and synchronization with the organizational/developmental system.

An inventory of tools

An example of a competency that must be developed systematically by individuals as well as the organization is the concept of inventory. The acquisition, storing, retrieving, and redistribution of inventory is costly to any industry, particularly the educational and information industries. Library orientations are designed to acquaint learners with the inventory of raw materials—theories, ideas, concepts, principles—of the education and information industries. To what extent do we introduce learners to inventories of data and information bases that are automated and stored in various repositories?

The concept of inventory is particularly important to the professional educator in vocational and technical programs. Federal legislation in 1954 and 1963 created a series of nine research and development centers to engage in R&D in education. In 1965 Public Law 89-10 created a network of 20 regional educational laboratories to translate the results of this R&D into products that could be used in education and training. Although the number of such laboratories has been reduced to 17 over the years, the primary function of this sytem continues to be the generation of new knowledge and the reduction of the lag between R&D and its dissemination and adoption. This system is coordinated, to some extent, through the Council for Educational Development and Research.

Another tool that has evolved over time is the system of Educational Resource Information Centers (ERIC). ERIC is a system of 17 clearinghouses that are repositories of documents ranging from pure through applied R&D for a variety of areas, such as adult, career, and vocational education; counseling and personnel services; educational management; higher education; information resources; junior colleges; reading and communications skills; and other topics. The value of these tools is apparent in the Spring 1982 issue of the ERIC *Junior College Resource Review,* "Literacy in Community College." Occupational education must utilize the inventory of tools to improve different types of literacy and improve the quality of the education and training industry.

Supervisor development

Equally if not more important to individual development, it is absolutely essential that institutional leadership understand the stages of adult development of the spirited professional they supervise. The concept of the supervisor as a rational and analytical problem solver with predominantly "organization man" characteristics is changing. Motivational psychologist David McClelland has noted that achievement was the motive of the sixties and power was the motive of the seventies. Daniel Yankelovich indicates that the managerial style of the eighties will be built upon human relations skills, characterized by a tough-minded as well as a warm-hearted approach.

This new approach to management in the eighties will require social skills. These new skills include communicating well at all levels of the organization; communicating expectations and standards; being open, honest, and fair with people; developing trust with people regardless of racial, sexual, or cultural factors; confronting problem situations and issues; developing skills in group and individual effectiveness; being a coach and mentor; and recognizing and molding the skills of others. Educated people do not like to be controlled from the top. Supervisor development, therefore, must focus on processes and skills that articulate organizational and individual issues.

ACCOUNTABILITY AND LEADERSHIP

In recent years Americans have become concerned about the return on investment in education generally and specifically in vocational education. The high cost of the labor intensive education industry, the unwillingness of the industry to change with the changing needs of society, the rate of obsolescence and high cost of equipment, and the cost-efficiency of proprietary occupational training causes taxpayers to ask, Is support of education worth the return on investment in terms of the individual and value added to society?

Several persons have commented on need for change in the educational system. Miller and Woock state, "The most immediately useful conclusion of the Coleman study is that the total environment of the school needs to change if its effectiveness is to be improved."[11] Kerr states, "The educational system

11. Harry Miller and Roger Woock, *Social Foundations of Urban Education* (Hinsdale, IL: Dryden Press, 1973), p. 348.

has contributed to the development of two distinct classes in America, an educated class and an uneducated class."[12] Drucker indicates that the demand for education is going up, not down, What is going down, and fairly fast, is demand for traditional education in traditional schools: The biggest infrastructure challenge for this country in the next decade is not the billions needed for railroads, highways, and energy. It is the American school system, from kindergarten through the Ph.D. program and the postgraduate education of adults. And it requires something far scarcer than money—thinking and risk taking.[13]

The leadership challenge

During recent years, occupational education has come under a series of attacks. The criticism provides stimulation to review three national studies: *Vocational Education and Training: Impact on Youth, National Longitudinal Study of 1972,* and *Project Talent.* In addition the Education Amendments of 1976 (Public Law 94-482) require the National Institute of Education (NIE) to transmit to the president and to Congress a study and evaluation of vocational education, the first of which is *The Planning Papers for the Vocational Educational Study, Vocational Education Study Publication No. 1.* The NIE was given the charge of conducting a "thorough evaluation and study" including "the distribution of occupational education funds in terms of services, occupations, target populations, enrollments and educational and governmental levels and what such distribution should be in order to meet the greatest human resource needs for the next ten years."[14] The introduction of vocational education into the public school system required adoption of a broader view of what public education should be. Many of the issues of vocational and technical education are becoming increasingly important.

Over the past several years Congress has tended to create new programs for the sake of major economic concerns, rather than to improve delivery systems already in place. The Comprehensive Employment and Training Act and the Private Industry Councils are but two of several such examples. As ocupational education has grown in importance, several public interest groups have become actively involved in seeking to influence the nature of vocational education reauthorization. Leaders in occupational education must be active participants in policy formulation on a broad range of issues, including economic revitalization.

An understanding of the policy process, as well as technical knowledge of substantive policy areas, is required of those who work in the public sector or seek to influence policy decisions. This requires a framework for translating social and scientific knowledge into public decisions and for understanding the effects of governmental policy. It requires a capacity to search out and appraise evidence from a variety of sources, to think critically about policy

12. Clark Kerr, *Giving Youth a Better Change: Options for Education, Work and Service* (San Francisco: Jossey-Bass, Inc., 1979).

13. Peter F. Drucker, "The Coming Changes in Our School System," *Wall Street Journal,* 3 Mar. 1981.

14. *The Planning Papers for the Vocational Education Study: Vocational Education Study Publication No. 1* (Washington, DC: National Institute of Education, 1979), p. v.

issues and policy proposals, and to make sophisticated use of the key theories, findings, and analyses that have been useful in policy formulation.

There are hundreds of commissions, committees, and task forces at work analyzing societal problems and submitting recommendations about courses of action. It is important to keep tabs on how commission members, feds, congressmen, and other policymakers size things up. We need to know their insights, perceptions, moods, and hunches. What they are thinking will affect what they will do. We need to go beyond the passive approach of keeping tabs on their thoughts, however, and be proactive in terms of legislator education. In the United States 7565 persons serve the 50 states as elected legislators.[15] For the most part they are citizen legislators who work most-time at various trades and professions in their home communities and part-time as lawmakers in state capitals. It is impossible for them to be knowledgeable about all issues without systematic cultivation. Their continuing education cannot be left to chance alone.

There is at least one two-year college in 426 of the 435 congressional districts. There are approximately 540 boards locally appointed or elected to govern two-year public colleges; in 17 states complete legal responsibility for the governance of public two-year colleges rests with a state-level board rather than a local governing board. It is difficult to relate the world of education to all policy areas, such as reindustrialization, unemployment, and economic development. It is equally complex to attempt to achieve anything reasonably near consensus between diverse governance structures and the profession when dealing with such policy issues as the federal role in vocational education, investment in research and development, or international markets.

CONCLUSION

The industrialized nations of the world are in the turbulent times of a structural shift from an industrial to a technological society. The scope and rate of change of science and technology is unprecedented, and its effects are cultural, psychological, social, and economic. The central question is how it will affect society and whether people will be the beneficiaries or victims of science and technology. What is needed is a conceptual framework to guide vocational and technical education in such a way that we focus science and technology on the individual and quality-of-life issues.

The evolution of the technetronic society can develop systematically if we can manage the issue of intellectual capital formation. Many of the old smokestack industries are gone forever. Economic recovery will be spread unevenly across the nation, attributable in part to such forces as the military build-up and defense and aerospace contracts. Economic and demographic forces are making the various sections of the country less, not more, alike. Many call robotics and other high-tech advances the panacea, but those of that persuasion should consider the location of Zenith, RCA, General Electric, Hewlett-Packard, Mattel, and Atari plants in Puerto Rico, Hong Kong, Taiwan, and Asia.

No issue will be more important in the next decades than the relationship of vocational and technical education to

15. *Special Report* (Princeton, NJ: Robert Wood Johnson Foundation, 1977).

the economy. In the past vocational-technical education saw its relation to the economy primarily in terms of providing a trained work force. This focus will continue to be important in the future, but it will not be sufficient. New expanded relationships will be required between occupational education and the economy in the computer-literate, high-technology, information society.

ERRATUM

Please note the following correction to Elmer N. Lear's review of *The Weimar Republic and Nazi Germany,* which appeared on pages 175-176 of the September 1983 volume of *The Annals.* On page 176, in the right-hand column, line 13 should read, "between Berlin and Warsaw," rather than "between Berlin and Moscow."

ANNALS, *AAPSS*, **470**, November 1983

Computer Aided Design and Computer Aided Manufacturing: The Wisconsin Experience

By JOHN G. BOLLINGER

ABSTRACT: Computing has become an integral part of an engineering education. From the moment a student arrives on the University of Wisconsin—Madison campus, until graduation with advanced degrees, the educational program provides convenient access to appropriate computing facilities to enhance a student's experience and expand learning opportunities. If engineering education is to contribute to future industrial growth and economic recovery, graduates must be well grounded in the principles and techniques of computer aided engineering in research, design, and manufacturing.

John G. Bollinger is dean of the College of Engineering, University of Wisconsin—Madison, and John Bascom Professor of Mechanical Engineering. He holds Ph.D. and B.S. degrees from the University of Wisconsin—Madison and an M.S. degree from Cornell University. He is active in the research and development of computer control systems, computer aided design, manufacturing, and robot intelligence. For the past 18 years he has conducted research in the identification and elimination of machinery and product noise.

THE computer has been revolutionizing our society for the past three decades. Computer management of data, scientific computation, computer supported machine intelligence, and interactive computer graphics have been major milestones affecting engineering practice, research, and education. Robotics, for example, is a subject derived from a combination of machine design, kinematics, computer implemented programmability, and machine intelligence. The opportunities that are emerging for enhancing national productivity and competitiveness in the manufacturing sector present one of the most stimulating intellectual challenges of the next decade. It is no longer possible for our society to isolate the scientific, technological, and sociological aspects of developments in moving ahead to maintain a strategic position in the world economy.

For over 20 years the University of Wisconsin—Madison College of Engineering has benefited from a strong tradition of developing the use of computers in engineering instructional problem solving and research development, and in real time control in machine intelligence. In recent years the needs for computing services, mini- and microcomputer acquisition, and the development of support staff have grown at a rate that has taxed the financial resources of the relatively large college. By 1981 the growing demand for a wide variety of computing equipment stimulated the development of a collegewide coordinated plan for computer services development that has now resulted in broadband access to a cafeteria of computing facilities and services, as well as a large number of local microprocessors and minicomputers in instructional and research laboratories.

The guiding philosophy behind the development of computing in the College of Engineering has been, and will continue to be, that in a modern engineering educational institution, all students, faculty, and staff must have convenient access to the computing resources they require to accomplish their work, to learn, and to advance the state of the art.

It is important to recognize that this philosophy applies to the ability not only to compute and to manage data, but to use the computer as a real time manipulator of machines and processes, and to communicate hierarchical structured information to responsible humans.

This article provides an overview of how one engineering college is advancing to achieve the goals developed to meet our technological and economic challenge of the future.

INSTRUCTIONAL COMPUTING

For the computer to evolve into an effective tool for designers, they must be educated to use it, to understand it, and to communicate with it. They must understand the traditional approaches to design and visualization, how mathematics can be used as a tool for design, and the precision of its calculations. At the university level it is not sufficient merely to train computer aided design (CAD) draftsmen; the computer must become an integral part of engineering problem solving and communication.

Initial experimentation with integrating computer graphics into the graphics education of preengineering students at the University of Wisconsin—Madison College of Engineering revealed that a hands-on approach to graphics via microcomputer is more valuable than a traditional lecture format. The educational advantages of the interactive, stand-alone, networked microcomputer are proving superior to those of canned programs.

An experimental course was developed by Professor James McNeary for first-year preengineering students, based on an existing introductory course in descriptive geometry. Extensive computer programming techniques are studiously avoided in the students' initial exposure to the computer. Instead they learn the rudiments necessary to plot first points, then horizontal and vertical line constructions, and basic mathematical functions as obtained from simple expressions. The integration of descriptive geometry and solid analytical geometry greatly enhances using the computer graphics approach. The stand-alone, locally networked microcomputer provides instant response and instructor interaction, not available with burdened time-sharing systems.

In an effort to integrate the use of the microcomputer in a basic preengineering class, room configurations and laboratory facilities of several institutions were observed. It was commonplace to have terminals and students facing the perimeters of the room, and any printing was done either at a remote site or by manually patching to a shared printer. This arrangement certainly did not represent the optimum.

The Wisconsin experiment designed and constructed a U-shaped classroom using 12 microcomputers, with all 24 student stations facing the center of the room. With this arrangement any material presented on the blackboard is readily observable, and the instructor has easy physical access to each student station.

The results of the Wisconsin experiment have been both rewarding and inspiring. The faculty associated with this effort are convinced that the format and techniques employed can produce a revolution in engineering classroom instruction, and they stand ready to advance these efforts through additional experimental laboratories, clear documentation for students and staff, and supporting video aids. As a result of the publicity accorded its success, numerous groups have been exposed to its unique characteristics—workshops involving architects and engineers, the entire College of Engineering faculty, high school and grade school students and teachers, and innumerable visitors from industry, both national and foreign. An indication that this laboratory has touched on some fundamental teaching issues is revealed by experiences with primary and secondary school teachers and students—every grade level from one to twelve. The teachers are always intrigued and excited about the classroom format and techniques. The students are enthusiastic about the notion of instant assistance, interaction with the instructor and classmates, and, perhaps most important, immediate recognition or reward when their solution is shown to the entire class.

In the upper-division levels, computing is playing an important role in engineering instruction in every department of the college. Access to a cafeteria of computers is provided by over 100 alphanumeric terminals and about a dozen limited capability graphics terminals. About 50 terminals are distributed between two major terminal access sites. Complete documentation and consulting services for all users are provided at the major sites. Other terminals are distributed throughout departments for instructional and research use. Terminals are connected to the appropriate computers under user control through a local area broad-band communication network. Work has already begun to extend this network to include a terminal in every faculty member's office. This project should be complete by the fall of 1984.

In October 1981 the Engineering Computing Committee unanimously endorsed the formation of a new organization and management structure for computing in the College of Engineering, titled the Computer Aided Engineering (CAE) Center directed by Professor John J. Uicker, Jr. The overall mission of the CAE Center is to provide up-to-date facilities and services, and to foster the intelligent use of CAE methods in the research and instructional programs of the College of Engineering.

The CAE Center is providing approximately 250,000 hours of interactive terminal access time per year to approximately 4000 active student and faculty users, at $1.60 per hour. In spite of the very rapid recent growth it is still obvious that even more computing resources are and will continue to be needed if engineers are to be trained appropriately in the latest technologies, such as CAD and computer aided manufacturing (CAM).

The total user community of the CAE Center numbers over 8000. However, their diversity of backgrounds and needs are even more significant than their number.

PHYSICAL FACILITIES

The CAE Center, as its title implies, integrates and centralizes the administration of college-funded computing. However, this does not imply centralization of the computer access or services provided. On the contrary it is assumed that computing resources can be best utilized only if the service is conveniently available in close proximity to the user. With the cooperation of the Madison Academic Computing Center and an appropriate communications network, CAE attempts to achieve both centralized management and staffing and distributed delivery of computing to its users.

Becoming involved in modern computing in any organization today requires space, equipment, operating funds, staff, and a strong commitment. In these days of precious resources it is not reasonable always to ask for more resources. Rather resources need to be reallocated to accommodate change. In the case at hand, no new space was acquired, and a creative, cooperative relation with other campus resources has resulted in tripling computer resources with only small increases in funding. Talented student help is used extensively, and a small, highly qualified operating staff provides excellent technical and managerial guidance.

ROBOTICS AND COMPUTER AIDED MANUFACTURING

A further dimension of the college program involves expanded emphasis on robotics and a broad range of interdisciplinary subjects related to CAM.

For over 20 years faculty in the College of Engineering have been involved in research and instructional efforts in robotics. Kinematics, automatic controls, instrumentation, and process physics form the basis for this work. Cutting across traditional department lines, faculty and students interact in interdisciplinary programs of study and research funded by individual companies or consortia. Until recently engineers have not recognized robotics as a separate intellectual subject. It was possible to advance the state of the art with a selection of existing subjects. Now it has become popular to compartmentalize those aspects of the traditional fields to form new combinations of pertinent information in the form of courses on robotics. For example robotic applications have become sufficiently diverse and plentiful to constitute a unit of study in itself. This can integrate the technology with the sociological and economic aspects, because they are no longer one-of-a-kind experiences. As in many other areas of industrial experience, the early innovators and investors have made enough mistakes to reveal the need for further insight.

Computer technology has been the key ingredient in major advances in robotic devices in the past decade. Computer generated intelligence and programmability, coupled with new dimensions of electromechanical sensory perception, give robotic devices the flexibility and performance capability required. Man's physical capabilities can be not only matched but exceeded in many functions.

The robot is but one aspect of the intellectual challenge that faces society in the production of goods. The computer holds the key for controlling and integrating all aspects of our manufacturing environment. Product quality is enhanced by increasing machine intelligence and providing on-line process sensing and control. Production reliability is enhanced by machine self-monitoring and diagnosis. Computer systems can integrate systems, optimize process equipment utilization, and minimize inventory. These results of computer managed and computer controlled factories of the future are all necessary ingredients in meeting the competitive challenge of the world marketplace.

RESEARCH

Research programs in the college account for a large fraction of computing resource use. Many of the programs have specific applications for CAD or CAM techniques. The major goal of many of the programs is the development of techniques that can improve productivity in an industrial setting. Examples of techniques being investigated include the following:

1. Optimal and adaptive sensing and control strategies for a variety of manufacturing mechanisms, including robots, drilling machines, die sinking machines, and other machine tools. Distributed computing techniques are being developed to allow efficient local sensing and control of mechanisms using dedicated processors. Links between the CAD host mainframe and the local processors are considered essential to integration of

the CAD and CAM functions through a CAD data base.

2. Interactive design techniques that improve designer productivity. Methods are being developed to aid in the design of very large-scale integration (VLSI) components, robot and machine mechanisms, cast metal parts, and buildings/structures. Methods are also being developed for the design of manufacturing plant layouts and production schedules. Distributed computer techniques are being investigated to support the design process more efficiently. There is a specific need in several programs to develop low-cost design work stations that small and medium-sized industrial users can afford. Again links between the local design processors and the CAD host mainframe are considered essential to allow development and use of large engineering data bases.

3. Interactive analysis and testing techniques that improve the productivity of the researcher and the practicing engineer. Methods are being developed to simplify analysis of power distribution systems, power plant failure modes, electromagnetic fields in electric machines, and the properties of manufacturing materials. Many of the design activities mentioned in item 2 also require development of efficient analysis techniques. Simulation techniques that take advantage of user interaction are also being developed in the area of biomedical engineering. Interactive testing techniques are being developed for application in the verification of VLSI component designs. The analysis calculations are generally computer intensive and are therefore best performed on a mainframe-sized host computer.

SUMMARY AND COMMENTS

Engineering education is undergoing significant change. To many it may seem as though the technology it is creating is overtaking the creator. Educators and practitioners alike are discovering that even avenues of endeavor that did not significantly interact with computers before are now becoming dependent on them. A modern engineering education requires not only computer literacy but computer intimacy. Engineering teaching and research laboratories require direct computer access, and research in many areas is focusing on advances made possible by computer technology.

These changes will alter the relationship between engineering and the social sciences. Decisions on the allocation of resources between manpower and computer power cannot be made without integrating all perspectives and projecting both short- and long-term results. The success of technological achievement brought about by interdisciplinary activity in the physical sciences and engineering must be matched by interaction between engineering and the social sciences.

Finally there is a more subtle problem developing in engineering education that may have serious long-term effects. The increasing costs of being in the business of computer-oriented engineering education is forcing an internal redistribution of financial resources to support new development. Faced with a shortage of new faculty, state-supported engineering colleges across the country are being forced to decrease enrollments. Technological education, which was so freely available to so many, is now being directed more toward the intellectually superior student. Under these conditions there may be less opportunity to interact with other disciplines. The possibility for biased judgments, relative to the relationships between society and technology, could become even greater, and our destiny may be in the hands of highly computer-oriented technical decision makers.

ANNALS, *AAPSS,* **470,** November 1983

Wanted: Managers for the Factory of the Future

By WICKHAM SKINNER

ABSTRACT: An obstacle to progress toward major technological change in American industry centers around the industrial manager and prevalent managerial processes. New technology promises a factory of the future with radically better and more challenging jobs and work environments but, in spite of these and major strategic advantages, the rate of investment and change is slow. Case research suggests that decisions to redesign and reequip factories with available new equipment are delayed by detail-centered, cautious managers enmeshed in typical corporate decision-making that focuses on finance rather than strategy. The investment costs and risks in new factory technology are substantially greater than present technology, augmented by the need for closely integrated equipment and information systems. Hence the start-up and debugging problems are formidable, the outcomes are nearly unpredictable, and the behavior of managers unwilling to assume such career risks is therefore altogether rational. This set of factors has outdated not only many present corporate investment decision-making processes but a heretofore successful generation of top industrial managers.

Wickham Skinner is the James E. Robison Professor of Business Administration at the Harvard Business School. Following an engineering degree from Yale University and army service, he received an M.B.A. from Harvard. He was employed by Honeywell (1948-58) in manufacturing, sales, and financial management before returning to Harvard for his doctorate. He has been teaching at Harvard since 1960, focusing on production and human resources. His books include Manufacturing in the Corporate Strategy.

AFTER two centuries of providing ever higher standards of living to Western civilization, the remarkable institution called the factory is now drifting toward failure. Its success in pouring out low-cost, high-volume production of manufactured products has produced economic benefits for every society that has invested in it. Today, however, many of our factories have become liabilities instead of assets.

Our factories are failing us primarily because they are increasingly noncompetitive. Productivity has seriously slowed down its heretofore inexorable climb. The output is often shoddy. New product introduction takes longer than in factory systems sited in several other societies. It is failing us too because it is an unhappy institution: its employee-members often speak of it as a jungle, stressful in its demands, tedious and repetitious in its job content, full of territorial and jurisdictional conflicts, governed via authoritarian processes of command and control, and unhealthy—often even dangerous—in its demands on the humans it employs.

The factory has been implicitly condemned as an unattractive institution by the millions of our young who, voting with their feet, turn down factory jobs if other means of earning a livelihood are possible. And to the high-potential men and women entering the job market from graduate schools of business administration, manufacturing management positions have ranked for 25 years near the bottom of alternative career paths.

Even its strongest advocates admit that in spite of its usual outstanding economics, the factory has never been a placid, contented institution. Its performance has been marked from the beginning by strife and discord between employees and managers, workers and owners.[1] From the early factories in England to the frequently violent, bloody battles of the 1930s and 1940s, unions of frustrated workers have bitterly contested "the rights of management to manage," sometimes even sabotaging the very machinery that provided their jobs. Factory jobs have been tolerated at best, hated at worst, and accepted only as a last resort for earning a living—a bad job accepted for a good wage. And from the late 1960s on, the changing sociology of expectations and demands for a higher quality of work life have served to increase employee discontent.[2]

These are not the marks of a successful social institution but of an institution showing signs of stress and obsolescence. Particularly sobering is its apparent inability to change itself. The auto and steel industries, for example, are victims of their own massive investments in the wrong equipment and process technologies (EPTs). Their dinosaurian mass production systems have been unable to change products flexibly and quickly enough to meet international competition in the late 1970s.

In a curious irony the factory, first spawned by the industrial revolution—a revolution of new mass production technology—is now being strangled by the very same kind of EPT, mass production, that made it so uniquely productive when it burst onto the scene 200

1. Andrew Ure, "The Philosophy of Manufacturers" (1835), in *The Philosophy of Manufacturers,* ed. Michael Brewster Folsom and Steven D. Lubar (Cambridge, MA: MIT Press, 1982), p. 365.

2. Wickham Skinner, "The Anachronistic Factory," *Harvard Business Review,* Jan.-Feb. 1971.

years ago. Today our mass production is a strategic millstone, dangerously inflexible in its ability to change products and output levels, while offering distasteful, monotonous jobs for a high proportion of employees.

So what is the future of the institution? Surely it will continue in some form, for the world's population wants ever more manufactured goods. Consumer demand, raw materials, EPTs, and energy are generally available, and in some societies people still seem to want to work in factories. Somewhere the factory will go on; but what is its future for Western society? Our interest naturally focuses on Western society and the U.S. factory. Our questions include, Will we reindustrialize? Will life in the factory become more attractive and rewarding? Can we regain, in some industries at least, a substantial competitive edge?

Now enters the factory of the future,[3] the ever elusive dream always just ahead, talked about since the early days of automation and robots in the 1950s. Since the "factory of the future" has many meanings, let me simply so label the constellation of new, highly mechanized, automatic, hands-off, low-labor content, self-correcting, self-controlling EPTs we have seen just ahead for 30 years and now are actually introducing. What will be the impact of this new technology on the future of the factory? Will it make it a more successful institution? A happier one?

In this article I will take the position that the answer is becoming clear enough to be beyond much debate: the new manufacturing technologies are so different in kind from what has been available in the past that their effects—when the investments are made and fully implemented—will be sweeping. They will above all allow our factory systems to be more flexible and responsive to the new industrial competition[4]—competition based largely on ever more rapid changes in product and process technology—and create largely different and improved human jobs and working environments.

Let me just go back to the hook slipped into the last paragraph—"when invested in and implemented." For just when this will come about is not so clear. The technology looks highly promising, but we have to invest in it and manage its implementation. How are these stages progressing? What have we learned so far? What is the role of managers or managements in bringing the factory of the future into being?

Some data bearing on these questions have emerged from my research in 12 major manufacturing firms in 1981-82. The article will review these findings, offering the conclusion that the new technology is indeed hopeful in its promise but is being delayed by present industrial managers—those key players who must be the movers and shakers in any industrial change scenario.

Managers, it will be argued, are the chief agents holding back these promising moves into new manufacturing technology. The roadblock is of course neither organized nor explicitly intentional. There is no conspiracy. Indus-

3. "The Factory of the Future: Always in the Future?" in *Towards the Factory of the Future,* ed. L. Kops (New York: American Society of Mechanical Engineers, 1980).

4. W. Abernathy, K. Clark, and A. Kantrow, "The New Industrial Competition," *Harvard Business Review,* Sept.-Oct. 1981.

trial managers are impeding change because of conventional, short-term modes of corporate decision making,[5] combined with the underlying premises and managerial characteristics of a professional breed of manager, nurtured and hardened by a 200-year history of self-reinforcement. In particular it appears that thinking patterns centered around that good called productivity may in fact be the most pervasive roadblock of all.

The article concludes with some encouraging evidence that a more effective paradigm of industrial objectives and decision making may now be emerging, driven in part by a new breed of industrial manager who appears to be emerging by a Darwinian process from the present scene of crisis and failure.

THE NEW MANUFACTURING TECHNOLOGY: PROGRESS AND PROBLEMS

In 1981-82 I visited a dozen large American manufacturing firms in a variety of industries asking the following questions: (1) What is the extent to which your firm has employed the new manufacturing technologies? (2) What has influenced the extent of the penetration and the effectiveness of these new EPTs in your facilities? (3) May I interview managers considered by top management as "best, high potential 'comers' who will probably be running major facilities in five to seven years?"

Some initial research findings

From these visits, interviews, questions, and subsequent analysis of data have developed the following preliminary research findings, expressed in necessarily sketchy and cursory assertions, as follows.

The new technology for sale is emerging thick and fast. It is being driven primarily by electronics and computer firms but also by industrial equipment firms, all of whom see large, even gigantic markets ahead. Foreign suppliers, especially in equipment, are playing a major role. U.S. machine toolmakers produce high-quality equipment but are risk aversive and fail to offer more than small chunks of a total system's needs. They tend to base their business planning on the objective of surviving in low periods and growing modestly in good times, and hence are not very aggressive.

The potential impact of the new technology is enormous. Its present penetration and impact, however, is very modest in all but the already largely mechanized process industries. It is moving slowly for many interrelated reasons. Some of the reasons include the fact that the technology is new and little understood, yet much feared, by potential users. A typical problem is that the new equipment does not work well for two to five years, and "those years are hell" for operations management. It is generally looked upon with skepticism by organized labor. Workers have mixed experiences in adjusting to it. Sometimes it is stressful to be in charge of so much expensive machinery, but often it makes the actual job far more interesting. The new technology substantially changes the jobs of supervisors and middle management, shifting the focus from watchdog and disciplinarian to planning, training, and communicating.

The new technology also shifts management power within companies, conferring the power of knowledge and

5. Robert Hayes and William Abernathy, "Managing Our Way to Economic Decline," *Harvard Business Review*, July-Aug. 1980.

capability upon the systems' designers, coordinators, and troubleshooters. Conventional line managers experience a loss of influence. It is difficult to measure results of the new manufacturing technology because accounting methods and processes are lagging behind manufacturing technology by at least 25 years, but it is very clear that the major benefits are not in cost savings. The major benefits do not come about until virtually the whole factory is reequipped and all systems are changed over. Islands of automation offer only modest benefits, but vendors sell only islands or chunks. The technology and process equipment are fragmented across a churning new industry. No vendor, like a Bechtel or a Turner in power plants and chemical plants, offers a whole turnkey system with guaranteed results. This is what is needed but no one can do it yet.

No one can do it yet because the technology is so immature and the investment and risks are seen as out of this world. Few companies want to go first. "Let the competition bloody their noses," is the normal stance. But the normal stance is also, "let us experiment a little," "let us buy a few robots," "let us have staff people so as to be sure to keep up with the state of the art." So there are many corporate staff people muddling around in this. They go to conferences and talk to each other and plead for money and experimentation, all with much frustration.

Division managers are not buying, however. They are sensible people, and they simply compare projected capital invested to what they have now; knowing that the technology will not work for two to five years and, if they perform well and stay out of trouble, they will be promoted and gone by then, they conclude that they would be foolish to be the ones to go all out for a whole new production system now. Further, company capital budgeting systems are good at handling expansions or cost savings or bottlenecks, but they are not good at investing in uncertainty and nebulous, though promising, payoffs that are difficult to quantify.[6] Few companies—in spite of Abernathy, Clark, Hays, Leonard, Sasser, Skinner, and Wheelright[7] and their pleas and formulae for developing a manufacturing strategy—do have a manufacturing stratcgy. As onc manufacturing vice-president put it, "So how can you plan for new technology with a four to five year payoff if you still plan only a year at a time?"

The research also confirms what is supposedly generally known, that manufacturing managers are typically conservative, risk-aversive people who prefer sequential to parallel uncertainties, with a low tolerance for ambiguity, and a strong distaste for any prospect of wholesale changes. Their paradigm is productivity and efficiency, and that mind-set accounts for many of the problems and sluggishness of change in manufacturing.[8]

6. Bela Gold, "CAM Sets New Roles for Production," *Harvard Business Review,* Nov.-Dec. 1982.

7. William Abernathy, *The Productivity Dilemma* (Cambridge, MA: MIT Press, 1982); Abernathy, Clark, and Kantrow, "New Industrial Competition"; Hayes and Abernathy, "Managing Our Way"; Frank S. Leonard and W. Earl Sasser, "The Incline of Quality," *Harvard Business Review,* Sept.-Oct. 1982; Wickham Skinner, *Manufacturing in the Corporate Strategy* (New York: John Wiley and Sons, 1978); Steven C. Wheelwright, "Integration of Manufacturing Strategy and Business Strategy," in *Manufacturing Trends in the 1980s* (New York: Booz, Allen, and Hamilton, 1981).

8. Wickham Skinner, "Productivity Is the Wrong Objective," *Wall Street Journal,* 16 Mar. 1982.

But there may be a new breed coming on fast, characterized by nonlinear and variegated career paths. Featuring information technology competences, team project leadership, thriving on change and ambiguity, nearly disrespectful of their elders, they are not as stubbornly loyal to their companies as were their predecessors. Their style contains a strong but sometimes unrealistic measure of self-confidence. They will be coming into substantial power in five to ten years, but few have it now.

The new manufacturing technologies will be powerful and influential, probably far more than most corporate executives realize. They are based on the microprocessor and ever cheaper computing and memory power. They will cause the replacement of much traditional labor. Their biggest advantage may be to create an order of magnitude increase in flexibility for product variation. They can economically produce very short runs. They will be self-correcting and improve quality and reliability. They will replace human variability with machine consistency.

Their biggest impact, however, will probably be strategic: to make manufacturing more readily a competitive weapon because of their speed in new product introduction, customer specials, short runs, and superior quality.[9]

The impact of the new technologies will be particularly powerful in their effects on working environments and human resource management. Many jobs will be less physical, monotonous, dirty, and dangerous. More jobs will require more knowledge and judgment. More jobs will focus more on keeping the system going and less on production output in units. Workers will have more autonomy, freedom, information, and discretion. Supervisors may become more coordinators, implementers, and sources of knowledge and expertise, and less the disciplinarian, counter, and cost watcher. The corporate name of the game will be more than ever human resource management, with corporate performance in recruiting, selection, training and development, and communications becoming critical, as the emphasis shifts to managing more capital-intensive and knowledge-dependent production systems.

While progress is still slow, this is after all the normal pace and pattern of major technological change. The power of the technologies, the coming of the new breed, and the phenomenon of one company in an industry gaining a major strategic, competitive advantage over the rest and thus setting off a grass fire that sweeps through the industry—all imply that in a five- to fifteen-year period profound and dramatic changes will occur in many production factories and service operations. These changes will enormously alter and indeed reinvent this old and slow-moving institution, the factory.

Industrial managers and their leadership

What is going on in this once great factory institution seems paradoxical. Why is the factory's social and leadership infrastructure holding back the very changes that might be its economic salvation? While the factory is apparently slowly dying in the Western world,

9. Wickham Skinner, "Getting Physical: New Strategic Leverage from Operations" (Working paper 83-43, Division of Research, Harvard Business School, Cambridge, MA).

its leadership seems unable to make the decisions, take the risks and invest the capital to reequip itself and survive.

So the question is, Why not? From the viewpoint of the observer outside the system, it appears to be illogical and shortsighted behavior; but to the actors on the scene their risk-aversive, business-as-usual response is not so illogical. They are reacting rationally in accordance with the reward/punishment characteristics of their corporations' management systems: they must play by the rules of prevailing corporate capital budgeting and financial management systems. Their experience has taught them that anything that can go wrong will go wrong, and that any overlooked detail can stop the production line. When that happens they are criticized; so they behave as they have been conditioned and constrained. Their motto is "productivity now."

The net result of this set of conditions and responses is that the very technology that might help manufacturing companies recover competitive position and gain strategic advantages is being adopted extremely slowly. The few situations, among the companies studied, in which firms have moved ahead faster have featured one or both of the following: (1) top management has had a clear, long-range vision of the advantages and indeed necessity of investing in advanced manufacturing technology, and has said, "Do it"; and/or (2) manufacturing executives have been so aggressive, broad in their vision, successful, secure, and influential at top levels that they have forcefully urged the investment and prevailed in being allowed to lead their organizations into the learning process.

The evidence is that it takes one or the other—or both—kinds of executive to move away from the let-someone-else-go-first syndrome that prevails. Further, the evidence is clear that if top management forces the new experimental technologies on less than enthusiastic and able production managers, the learning process is inevitably more painful, expensive, and drawn out in time than when the production people are eager, ready, and proactive in their attitudes toward the new EPTs. This may appear only too obvious to the researcher, but the problem is that these management phenomena form a major roadblock holding back the industrial catch-up so badly needed in U.S. industry.

The weakness in production management that manifests itself in a cautious, careful, and narrow focus in relation to investments in the new manufacturing technology is not a new phenomenon. It has been evident in case studies of production management over the past 30 years. The production management field is characterized by slow progress and modest changes when compared to the rate of change in financial management, marketing, control systems, information systems, strategic planning, and human resource management. In each of these sectors of administrative science, progress in the past 20 years has been remarkable. We have seen the emergence of such concepts and techniques as present-value analysis, scientific market research and analysis, corporate budgeting and control systems, and totally new approaches in information system design and handling. Over 200 documented, experimental approaches in human resource management have developed, including nonsupervised work groups, totally new work restructuring and compensation systems, and most recently human resource strategic planning concepts and practices.

Advances in conceptual management in production must be considered modest. The much publicized materials requirements planning (MRP) is little more than simply a computerized process of ordering needed materials, which is what has been done for 40 years, only by slow and cumbersome hand methods. Statistical quality control was invented in the decade before World War II and then used less in the 1970s than the 1950s; we have added the advantages of project and program management learned from aerospace industries in the 1950s and 1960s. Operations research has offered the use of mathematical models for scheduling and inventory management but has had a disappointingly small impact in practice. One new management concept and practice area has been that of manufacturing strategy,[10] first conceptualized by John G. McLean in 1950, and even that concept has been adopted very slowly.

The ironic fact is that while scientific management began in the factory, factory management has been outrun by every other functional area of management since the halcyon days of Frederick Taylor, Lilian Gilbreth, and efficiency experts in time and motion study. Using a product or technology analogy, the field has been mature for a long time. Its rate of change and its members' rate of new learning has been relatively flat.

It is necessary to speculate on the reasons for the maturing of the field of production management by the end of World War II, its subsequently slow rate of change, and its present role in impeding progress toward the ever elusive factory of the future. We seem to have a generation of men—there are a few women but not many—who, compared to their counterparts in finance, marketing, control, and human resource management, appear to be cautious, reactive, and occasionally unimaginative. They are seldom promoted to top management. They are often criticized by top managers as being too narrow to see and understand the whole business, and as having an excessively tactical rather than a strategic point of view; they are functionally oriented rather than able to integrate well across other functions of the business. They are often seen by human resource managers as authoritative, dominating, and mechanistic in their personal management styles. In practice they emphasize details of costs, schedules, efficiency, logistics, set-up and running times, and productivity. Driven simultaneously by cost, time, delivery, and quality objectives that represent unrecognized trade-offs,[11] their factories are often unfocused[12] and anachronistic. When push comes to shove, as it has in the 1980s, the ingrained instinctive response is as always[13] to return to the productivity paradigm.

These men are the mud-slogging infantry of our industrial time. They are heroes in the sense that they dig in, do the nasty, detailed, day-by-day plan-

10. *Advanced Production Problems* (Homewood, IL: Irwin, 1952).

11. Wickham Skinner, "Manufacturing—Missing Link in Corporate Strategy," *Harvard Business Review,* May-June 1969.

12. Wickham Skinner, "The Focused Factory," *Harvard Business Review,* May-June 1974.

13. Alfred D. Chandler, Jr., *The Visible Hand* (Cambridge, MA: Harvard University Press, 1977).

ning, expediting, and follow-through necessary to get out the production and keep cost variances down. They are under relentless presure from above on conflicting performance yardsticks and face strident social and often union-driven pressure from below to keep discontented, often unhappy people working productively. They are a product of those pressures and their heritage. Their survival instincts create a kind of technological imperative that is a powerful influence. It is small wonder, therefore, that they behave as they do.

WANTED: A NEW BREED OF MANAGERS

What kind of managers and managements are needed to become positive rather than negative factors in bringing the factory of the future effectively to bear on our industrial problems? The research described before leads the analyst to the following prescriptive characteristics for the men and women who will be needed.

1. Knowledge will include comfort-level knowledge of mechanical, electronic, and management science technology, and all business functions—production, marketing, finance, control, personnel.

2. Skills will include system design—logistical, information, human systems, planning, coordination; project management and team building; interpersonal communications, handling large groups, and labor relations; general management, for example, integration of business and technical functions—in contrast to specialized technical skills.

3. Attitudes will be positive toward change, restless unless making progress in change, and will reflect security in professional life.

4. Cognitive/thinking style will involve the ability to conceptualize combined with the ability to work from specifics in a systematic and analytical rather than purely intuitive pattern.

5. Premises/assumptions will be that the objectives of a manufacturing system are multidimensional. Rather than always focusing the production system primarily on cost and efficiency, the manager must set objectives based on the strategic needs of the corporation. These objectives will include criteria concerning the following, specifically establishing both priorities among and specific trade-offs between:

—cost-efficiency-productivity;

—customer response time, that is, delivery lead times;

—reliability of delivery promises;

—quality of product;

—investment in equipment and inventory;

—flexibility for product changes; and

—flexibility for volume changes.

This is admittedly a pie-in-the-sky description of the compleat manager, who will bring the factory of the future into being. A contrasting description of the typical present-day manager also emerged from this set of research interviews. This is a composite picture that no one manager would completely fulfill, admittedly overdrawn to make the point:

1. Knowledge includes specialized, technical expertise generally limited to the existing product and EPT, and is usually heavily weighted toward one, either professional or functional, sector of production, such as production control, inventory control, or manufacturing engineering.

2. Skills include strengths in handling and dealing with large numbers of people, in expediting meeting schedules, and in cost control, with weaknesses in finance, marketing, and business strategy.

3. Attitudes reflect the assumption that management has a right to manage, and that I must keep control, using the authority conferred by the organization chart.

4. Cognitive/thinking style will in- on specifics and details, using intuitive analysis based on experience.

5. Premises/assumptions are that the manager's job is to command and control, and that productivity is the most important, overriding criterion for manufacturing success.

The contrast between the people presently in power and what appears to be needed is striking. The limitations of many production managers can be traced in part to three principal sources: (1) the industrial revolution and its legacies; (2) Frederick Taylor and the principles of scientific management; and (3) the influences of the industrial engineering profession.

The industrial revolution substituted steam and water power, combined with new, mechanically based machinery, for human power and craft skills. Its theme was the mass production of cheap goods. Its workers received wages that were extremely attractive. Its owners' returns on capital invested at risk depended on the phenomenon of heretofore undreamed-of efficiency. For the owners and managers the success of the operation depended primarily upon the achievement of efficiency.

A century later Frederick Taylor showed production managers that by planning, measuring, and systematic organizing, large increases in efficiency could be obtained. He was the pioneer industrial engineer and was followed by a long procession of superb analytical industrial engineers and efficiency experts. Their expertise was in detailed analysis of human and logistical movements, the establishment of stopwatch standards, time and motion study, and ultimately whole books of predetermined standards based on human physiology, which resulted in accounting standards and variance controls. It built year by year into a profession and established a pattern of thinking, the paradigm called productivity.

Its thinking dominates the scene even today. To be productive demands a systematic mental focus on details, time, schedules, dimensions, and minute operations. But being productive this week and this month is nearly impossible if destablizing changes are being introduced. Production managers seek the maximum of continuity of schedules, products, and processes. Productivity is a good that is hard to dispute. For years the more productive departments, factories, companies, industries, and countries tended to win out in competition. The problem today is that competition is based not only on productivity but on quality and delivery and flexibility as well. The achievement of quality and flexibility of volume and product in Western society depends upon a firm's ability to attract and hold outstanding people in its production operations.

The vicious circle is apparent. Thinking based on productivity dominates the industrial scene and drives out more practical and realistic premises that might include the multidimensional parameters needed; it holds back investment in and successful implementation of the promise of new manufacturing technology; it discourages the selection and development of men and women who might think of the factory in broad and more flexible alternatives, handle change and ambiguity, solve problems without resort to authority, and attract and develop more of their own kind. In the language of cognitive styles the four-way matrix in Figure 1 suggests some dimensions of the contrast between what generally predominates today and what appears to be needed for the factory of the future.

The emerging production manager—a new breed?

During the course of the research described earlier, I interviewed many comers, typically younger managers emerging as high-potential people who were expected by their firm's top managers to "be running the place" in five to seven years. The interviews reviewed the manager's education, training, and career path; open-ended questions also explored the manager's premises and attitudes concerning executive practices, problem solving, introducing changes, and managing people.

The data suggest that it may be possible that the managers needed are in fact emerging via the competitive processes of self-selection, survival, and promotion. The sufficiency of their numbers and the rate by which they will emerge is of course still open to question; but their qualities and potential for growth and meeting the challenges appear impressive.

Instead of coming up through the conventional career paths—relatively few but long-duration positions in first-line supervision, production and industrial engineering, or production staff departments—these men and women are apt to have had a greater variety of career paths. They have come from such work experience as research, design engineering, accounting, sales, marketing, or personnel. They often have had computer programming experience, been in charge of human resource experiments, installed small sections of newly automated equipment or experience with new plant planning and operations or program/project management.

Their positions are apt to have been of shorter duration than their predecessors. Their skills feature the abilities to form up and lead effective teams for problem solving, systems design, and experimental manufacturing systems. They are apparently rapid learners and tend to surround themselves with others who adapt quickly. They seem to thrive on change, uncertainty, and ambiguity and indeed become easily bored with routine production. They delegate easily and in fact rather loosely, relying more on trust and less on formal controls and reports. Their casual styles, somewhat surprisingly perhaps, seem to be acceptable and even refreshing to union stewards and business agents. They seem less loyal to their employers than to their professions, and might be termed "universal production system designers and managers."

The emergence of this new breed of production managers is not surprising as a response to economic and technological needs. What is unexpected is their success in what might have been predicted to be an alien environment. For these new breed managers are of course being selected and promoted by their older superiors, who represent an earlier and now declining elite. The old are developing the young not in their own image but in the new skills, prem-

FIGURE 1
COGNITIVE STYLES

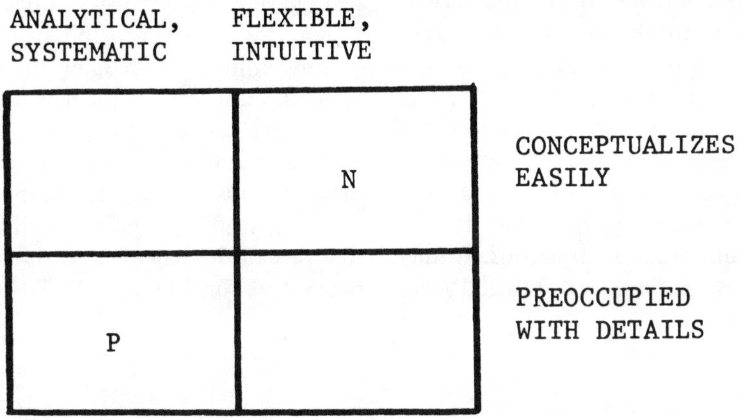

P = predominating today
N = needed for the factory of the future

ises, and attitudes that appear necessary and functional for today. It is as if the older generations recognize that what is needed now is not more of what they, the older generation, have had to offer, but something new.

CONCLUSION

Recommendations from this research into the management implications of our slow progress in investing in and implementing new manufacturing technology fall into two categories: recommendations (1) to corporate managements and (2) to thoughtful citizens and students of the industrial scene.

To managements, the scenario of slow progress due to managerial behavior molded by outdated business practices in capital budgeting, technological assessments, and management reward systems might ring out a warning bell of alarm. Sophisticated managers and students of administration have seen this same syndrome many times before, as have students of governmental and social systems. The decision makers become prisoners of their own systems of governance. This is what is happening in industry today. Industry needs to attack its own confining management infrastructure systems, and when these mechanisms are effectively overhauled, better decisions made with a boldness and daring appropriate to the times can be expected to begin remaking the factory.

The human resource system for attracting, developing, and nurturing a wholly new breed of production managers needs particular attention. An able new breed is developing, largely by selection of the fittest, but this process could be speeded up enormously by more deliberate, purposeful management development.

To the thoughtful observer outside of industry and corporate responsibility the implications are equally substantial. What is going on now, albeit slowly, is the reinvention of this old institution, the factory. Its direction suggests that life for people in future factories may be quite different from the bitter experiences of its past unhappy history. It suggests too that there are already exciting jobs in the factory for the able of our young, and great opportunities for young managers who seek careers in which they can make a contribution and work with people rather than dealing with second-order abstractions. It may be true now that in contrast to 200 years of mass production, the new microprocessor-based equipment and process technologies are positive engines of change moving industry toward the human uses of human beings in industry.

The limiting ingredient to a more rapid development of the new factory is no longer the technology or its economics; it is the manager and management systems. Thoughtful, responsible, broad-gauged, innovative and daring—these are the managers needed now for a new future for the factory.

ANNALS, *AAPSS*, **470**, November 1983

Technology, Competition, and Employment

By HARVEY BROOKS

ABSTRACT: Technology makes possible the production of existing goods and services with fewer input resources and the creation of new products and services fulfilling hitherto unmet or unrealized social needs. The accomplishment of both of these functions is difficult to estimate quantitatively, and it is thus hard to say anything definitive about the effects of technology on employment or on the quality of jobs, especially since technology is only one among several factors that influence this. Forecasts of future technological trends and their social impacts tend to overestimate changes in the short term, and underestimate them in the long term. This article stresses the key role of social and organizational learning, and the necessity for a higher level of cooperation among labor, management, and government in the selection and introduction of new technology and its accommodation to society. In the future the factory and the office must become more like a research laboratory or an advanced school where cooperative learning is the norm.

Harvey Brooks is professor of technology and public policy at Harvard's School of Government, and chairman of its advisory committee for the Program on Technology, Public Policy, and Human Development. He was dean of engineering and applied physics at Harvard from 1957 to 1975. A theoretical physicist, he has served on several government committees, including the President's Science Advisory Committee and the National Science Board. He is a member of both the National Academy of Sciences and of Engineering.

THERE are two principal functions of technology: to improve efficiency in the production of existing products and services, and to produce new products and services to fill better some social need or to meet a need not previously met or perceived. The first of these functions is measured by productivity, defined by the quantity of output in relation to all inputs—labor, materials and energy. For the purposes of these definitions production can be either in the private or public sector, and outputs may include public or collective goods such as environmental quality, public health, or physical security.

SOCIAL FUNCTIONS OF TECHNOLOGY

Both functions are more subtle and difficult to define in hard terms than conventional wisdom would suggest. For example, how is labor input to be measured in quantifying productivity? Conventionally this is done in terms of man-hours of direct and indirect labor without reference either to the skills and training embodied in a man-hour of work, or to whether a man-hour of work is dirty, or boring, or hazardous on the one hand, or clean, interesting, and safe on the other. A man-hour that embodies lengthy professional training and extensive learning through work experience ought to count as a more intense labor input than one that does not, since social resources are consumed in providing trained and experienced people. Similarly a social cost is embodied in a man-hour spent in a job that is dirty, unpleasant, boring, or hazardous, and this social cost should somehow be counted in measuring labor input. In other words there is a quality factor in a man-hour of labor input that should reflect both the quality of the job from the subjective standpoint of the worker and the social investment in competence development embodied in the worker's skills.

In most discussions of productivity and labor inputs, however, jobs are treated as fungible, and changes in quality or skill requirements of jobs are generally omitted in discussions of labor displacement due to technological change or foreign or regional competition. I know of no precise way of treating these factors; I can only point to them as a serious conceptual puzzle in the discussion of productivity and technological change. Is there an increase in productivity, for example, when an unpleasant, hazardous, low-paid job is replaced by a more pleasant one, but one that also requires much higher social input in terms of training and experience?

Similar quality questions arise with respect to other factor inputs used in measuring total factor productivity, such as materials or energy. What happens if a material input is substituted by another based on much cheaper raw material but with a higher value added in processing, so that the unit cost is the same? There would appear to be a social gain here, but one not reflected in conventional factor productivity measures.

In many cases productivity is measured as labor productivity, and the economic benefits of technology are thought of in terms of labor saving, or reduced labor costs. From the standpoint of competitiveness, however, savings in other inputs due to the application of new technology may be equally or more important than labor saving. Modern information technologies could turn out in many cases to be much more important for the savings they effect in materials use—through more refined design procedures—or energy use—through more discriminating direction of energy to the point of use—or for increasing the percentage of time that expensive capital is actually used in production. These savings of other factors become more important as these factors become relatively more expensive in comparison to labor, as they have in the last decade.[1]

If input measures are uncertain in estimating productivity, output measures are even more so. It is frequently said of the Japanese auto industry that the introduction of robots is more important for the improvement of product quality and the saving of inspection and rework than for direct savings in labor costs. The use of robots to replace stressful jobs may be particularly important in raising product quality, since job stress leads to deterioration in performance. Measuring the social value of the output also raises conceptual problems in estimating productivity gains, since this social value is usually taken as the market price for lack of a better method of quantification. Market prices may be very sensitive to income distribution, however, and many goods are produced in society that are not sold in the market.[2]

Even the distinction between technology that raises productivity and technology that improves products may be quite arbitrary. Consider, for example, a new technology that greatly increases the durability of a product. The same service to society may be provided by much less product output, since products do not have to be replaced as frequently. Under these conditions there could be as much labor displacement as would be caused by a more conventional productivity increase. The measure of productivity would depend upon whether output was measured in terms of a social service or a physical product. The high reliability of modern information technology is almost certain to prolong the life of many products through the replacement of mechanical functions by electronic functions, and this will have labor displacement effects just as important or more important than automation.

On the other hand, information technology is also accelerating the rate of improvement of many products, so that existing products—and services—may become obsolete more quickly, thus leading to increased demand. It is almost impossible to decide which will be the more important effect.

LABOR DISPLACEMENT

It is misleading to look at the effects of increasing productivity due to technology only within the context of single firms or single industries. The effects of technology depend on what happens to employment in the whole economy, not just the industry immediately affected. For example, if labor cost savings in inputs are reflected in the consumer prices of products, these savings result in the release of purchasing power, which in turn will result in demand and hence employment elsewhere in the economy.[3] The relevant employment-related ques-

1. Kathleen Newland, *Productivity: The New Economic Context,* Worldwatch Paper no. 49 (Washington, DC: Worldwatch Institute, 1982).

2. Harvey Brooks, "Technology, Evolution, and Purpose," in *Modern Technology, Problem or Opportunity?* ed. S. R. Graubard, *Daedalus,* Winter 1980.

3. OECD, *Science and Technology Policy for the 1980's* (Paris: OECD, 1981), esp. pt. 3, "Science, Technology, and Society in the 1980's."

tion is not how many jobs are lost in just the firm or industry experiencing productivity increase, but whether the resulting released purchasing power creates fewer or more jobs than those lost, and whether the new jobs created are better or worse than those lost.

Even within a single firm, whether productivity increase results in labor displacement depends on the elasticity of demand for the firm's output. If this elasticity is high enough, then the additional demand due to cheaper output will more than offset the lower labor requirement per unit output in its net effect on employment. The work of John W. Kendrick on the information industries has shown that rapid increases in productivity have been accompanied by a net increase in demand for all factor inputs because of the increased sales of the product.[4]

Some people have argued that although the situation described by Kendrick may have been true in the 1950s and 1960s, the increased mobility of capital, technology, and information coming about in the 1970s and 1980s in consequence of the spread of information technology has completely altered the picture. As soon as a new capital-intensive production technology is standardized it is a candidate for export to low-wage countries, where lower labor productivity is more than offset by lower wages.[5] Thus, it is argued, the globalization of competition, combined with large regional and national disparities in wage levels, more than offsets the effects of increased demand on output levels in the country or region of origin. In this more sophisticated view labor displacement is not a simple consequence of productivity increase due to automation, but of the combination of this with global competition, the ever-increasing mobility of capital and technology, and the increased span of management coordination made possible by global communications and transportation.

The validity of this argument would depend strongly on the importance of released purchasing power in creating demand in other sectors of the economy, and the capacity of these sectors to attract the necessary investment and of labor to move into them without undue adjustment problems. This would apply particularly if the new demand is created primarily in sectors less subject to the global competition effects operating in the industry from which the workers were displaced. Since indirect employment effects in the economy outside the firms that experience increases in productivity depend in a very complex way on the changing structure of both the national and the international economy, it seems very difficult to make meaningful generalizations about the effects of technological change on either the quantity or the quality of jobs. Some further observations on this, however, will emerge in the next section.

TRADED VERSUS NONTRADED GOODS AND SERVICES[6]

Traded goods are goods that are subject to both national and regional com-

4. "Impacts of Rapid Technological Change in the U.S. Business Economy and in the Communications, Electronic Equipment and Semiconductor Industry Groups," in *Microelectronics, Productivity, and Employment,* ICCP report no. 5 (Paris: OECD, 1981).

5. Telesis Consultancy Group, *A Review of Industrial Policy,* National Economic and Social Council, no. 64 (Dublin: NESC, 1982), esp. p. 44.

6. The discussion in this section draws heavily on ibid., chap. 1, pp. 44-63.

petition in accordance with factor costs and output prices, as well as such characteristics as product quality and performance. It is traded goods that are subject to competition from low-wage countries and regions, because of the growing mobility or transferability of the factors of production. Even the differences in skill level between high- and low-wage regions can be reduced by packaged training programs, whose availability is likely to increase and whose cost is likely to drop dramatically in the future. For traded goods, and services, improvements in productivity can result in the higher value added per worker necessary for raising living standards only if two conditions are satisfied:

1. The rate of gain of factor productivity continues to exceed that in competitive countries or regions because of input factors that are relatively scarce and immobile due to complex linkage effects among related technologies or institutions.

2. There are other high-quality jobs appearing for displaced workers elsewhere within the economy, most frequently as a result of continual development of new and more advanced products and services that are marketed in advance of the corresponding developments by competitors.

For nontraded goods and services living standards of the workers will be more dependent on absolute productivity levels and less vulnerable to low-wage competition. There is, however, an important interaction between the traded and nontraded sectors, because nontraded goods and services are input factors for the traded sector and therefore affect its costs and hence its competitiveness. For example, the wages of construction labor can have a big impact on the cost of capital facilities that require a great deal of field construction or assembly. Also labor costs in the traded and nontraded sectors interact through local, regional, or national labor markets, thus influencing labor costs in both sectors. For example, if labor or business can use a monopolistic position in a nontraded sector to raise costs, competition in the labor market may raise labor costs in the traded sector as well. Perhaps even more important, because of advances in information and communications, there is a tendency for what were once nontraded goods and services to become susceptible to interregional or international competition, and thus in effect to enter the traded sector. We now even have international companies offering various sorts of urban services to cities around the world. Similar developments have occurred in management services, professional services, construction, research and development (R&D), and so on. This may have the effect of gradually equalizing many factor costs both regionally and internationally, thus shifting the burden of maintaining regional and national competitiveness in traded industries more and more to strategic linkages between different parts of the business, such as R&D and marketing, or to effects of aggregation of different interdependent activities in a locality, such as higher education and high-technology innovation.[7]

THE PROCESS OF INNOVATION AND TECHNOLOGICAL PREDICTION

In trying to anticipate trends in employment and the quality and distribution of jobs from foreseeable devel-

7. Nancy S. Dorfman, *Massachusetts' High Technology Boom in Perspective,* report 82-2 (Cambridge, MA: Center for Policy Alternatives, MIT, 1982).

opments in technology, one faces a difficulty that seems to be inherent in technological forecasting. This is the tendency to gross overoptimism in the short term about the benefits or workability of new technology. This overoptimism also often leads to excessive expectations about the rate of adoption of new production technologies and, hence, exaggerated fears about labor displacement. What it takes to make a new technology reliable and predictable in performance is usually greatly underestimated, most especially by its proponents and entrepreneurs. How to use a new technology in the work setting, and to assimilate it into a working organization, is usually even more underestimated. For example, numerous case studies of the introduction of robots into factories and office automation into offices have consistently shown failure to realize the productivity gains that the proponents of the technology enthusiastically believed in and projected.[8] Underestimation of training needs, failure to recognize organizational requirements, increased overhead for monitoring workers, constant breakdowns and repairs of equipment have almost invariably accompanied the early stages of introduction. This has been most recently documented in the case of the still emerging information technologies, but it has been characteristic of virtually every major innovation in recent history. It has been shown, for example, that Japanese success in penetrating the U.S. subcompact car and consumer electronics markets owes a great deal to Japanese persistence in the face of initial disappointment and failure, as compared with their American counterpart firms in the same market segments.[9]

At the same time, however, the long-run impact of a new technology is frequently underestimated at least as badly as its immediate impacts or benefits are overestimated. The crises and trauma created by the early troubles often prevent people from giving adequate consideration to longer-term impacts and planning for them. Frequently also the long-term impacts turn out to be different from those anticipated when the technology first appeared on the scene. There is a tendency to exaggerate the possibility of foresight about social impacts, and in the past at least this has often been left to experts in the particular technologies involved, usually people with little sense of social, political, and psychological factors, and even less sense of history or cultural trends. The net result is thus often either discouragement and abandonment of innovative ideas to later competitors, or dire predictions of adverse social impacts followed by unwarranted complacency when these impacts do not appear instantly.

Another reason for underestimation of longer-range trends is failure to foresee the implications of scale. A new technology applied initially on a small scale in a few locations produces observable effects from which we extrapolate to the potential impact of a much larger scale of application or universal adoption of the technology. But there are systemic interactions and linkages through a whole economy or society

8. Joel A. Fadem, "Automation and Work Design in the Unites States," in *International Comparative Study on Automation and Work Design,* ed. F. Butera and J. Thurman (New York: North-Holland, forthcoming).

9. R. Rosenbloom and W. Abernathy, "The Climate for Innovation in Industry: The Role of Management Attitudes and Practices in Consumer Electronics," *Research Policy,* 11:209-25 (Nov. 1982).

that only appear in a nonlinear manner with increasing scale. These nonlinear systemic effects cannot be foreseen by simple extrapolation from initial experience.[10]

The phenomenon of short-range overestimation and long-range underestimation of the impacts of new technology leads to both exaggerated expectations of benefits and exaggerated fears, but when neither of these exaggerated expectations is realized a period of complacency or disillusion tends to set in, which discourages further attempts to look ahead and plan for the future. Thus unexpected successes and unexpected problems often burst on the world unheralded and unplanned for long after early discussion and debate on the implications of the technology have been forgotten. For example, fears of unemployment due to the spread of computer technology and automation led to the formation of the Automation Commission in 1962, but the fears expressed then proved unfounded as unemployment actually declined, and the commission announced that automation was a nonproblem as long as macroeconomic policy was appropriately managed. Subsequently the whole subject of labor displacement became unfashionable and was discredited for nearly two decades. Yet we can now ask whether the fears expressed in 1962 were merely premature, and whether we are now about to realize some of the dire predictions made then. As I have already indicated, I for one cannot answer this, precisely because of the complex linkages I have described. Although the situation certainly does give cause for concern, the impact of these technologies may not be inevitable but may depend on many concomitant nontechnical factors.

10. National Academy of Sciences, *Technology: Processes of Assessment and Choice,* (Washington, DC: Committee on Science and Astronautics, U.S. House of Representatives, July 1969).

One can also ask more broadly whether we can as a society find a way of breaking this unstable oscillation between exaggerated expectations followed by failures of foresight.

THE IMPORTANCE OF ORGANIZATIONAL LEARNING

As already indicated, many case studies of the introduction of new technologies into offices and factories demonstrate the importance of human factors and organizational considerations to the actual realization of the projected benefits of the technology. Frequently the newer technologies demand a higher level of cooperation, both among the members of the work force and between the workers and management, than was true for predecessor systems. With the new information technologies, for example, the possibilities for degrading productivity through unmonitorable or even only half-conscious noncooperation may be much greater than with mechanical production systems. By the same token there may be an equally great potential for unexpected productivity and quality benefits through effective cooperation and continuous organizational learning, in conjunction with highly flexible technological systems whose adaptability to different organizational designs can be achieved by software changes rather than hardware redesign. In recent experience less than expected gains have been the rule for new information technology, and for other technologies as well, but there are a sufficient number of isolated counterexamples to suggest that this disappointing experience is not a property of the technology itself.

Thus it appears that rapid technological progress in production systems, whether in offices or factories, will demand a matching pace of organizational and managerial innovation, which may entail rather fundamental changes in management philosophy. This may be feasible only with a highly participative and cooperative process not readily subject to central masterminding. Furthermore the capacity for such organizational innovation may become a key competitive advantage for firms or even whole geographical regions, which may determine where economic growth resulting from new technology will be located in the future—both regionally within nations and internationally. This kind of organizational capacity for adaptation and learning may be much less subject to technology transfer than hardware and software, and may thus be less vulnerable to low-wage competition than high-tech activities by themselves. To put the matter another way, localities or communities that first learn how to generate through higher levels of cooperation a self-sustaining process of organizational adaptation and innovation may enjoy large and relatively invulnerable competitive advantages in the market.

DIVISION OF LABOR

It is conventional wisdom that the competitiveness of mass production industries in the twentieth century has been based on the finer and finer division of labor. Short-term pressures for cost reduction in highly competitive businesses, as well as in many nonprofit or profit-making services such as health care and banking, are often met by further subdivision of jobs, deskilling many tasks while upgrading a few others. This leads to an increasingly differentiated, rigid job structure—a work organization that might be quite competitive under conditions of static competition, with slowly changing technology and markets, but which adapts very slowly and incompletely to the rapidly changing technology and markets we face today in many industries.[11]

It is urgent that as a society we move toward a different philosophy of the division of labor, one that pays more attention to the continuous upgrading of human resources through constantly enhancing and diversifying the skill repertoires of the entire work force, from top to bottom. This implies not only a system of lifetime learning within firms, but also the organization of work so as to promote learning and skill enhancement as an inherent part of the production process itself. The factory or the office becomes more like a research laboratory or an advanced school. This implies a pay structure based more on acquired skills and experience and less on precise job descriptions or formal educational credentials. Furthermore a work force within the firm that is continuously educated for adaptability will also be much more employable elsewhere in the economy, should its members be displaced by technology or competitive forces.

11. Charles Sabel, *Work and Politics* (Cambridge, England: Cambridge University Press, 1982), esp. chap. 5.

ANNALS, *AAPSS*, **470**, November 1983

Flexible Manufacturing Systems: Robotics and Computerized Automation

By M. EUGENE MERCHANT

ABSTRACT: The most powerful long-term technological trend impinging on the factory of the future is that toward computer integrated manufacturing. Behind this trend lies the unique capability of the computer to automate, optimize, and integrate the operations of the total system of manufacturing. The virility of this trend is attested to by technological forecasts made over the past 10 years. The rapidity of the development is due not only to this technological virility, but also to powerful long-term economic and social forces impinging on manufacturing. As a result many industrialized nations are pursuing large national programs of research, development, and implementation of computer integrated manufacturing to hasten the technological evolution of computer integrated automatic factories. Programs receiving major emphasis include development and application of integrated manufacturing software systems, group technology and cellular manufacturing, computer control of manufacturing processes and equipment, computer-controlled robots, flexible manufacturing systems, and prototype computer automated factories. This evolution poses some significant challenges to American industry.

M. Eugene Merchant is director, advanced manufacturing research, at Metcut Research Associates Inc., Cincinnati, Ohio. He received his B.S. in mechanical engineering at the University of Vermont (1936), and his Ph.D. with majors in physics, chemistry, and mathematics, at the University of Cincinnati under a fellowship sponsored by Cincinnati Milacron. He is a member of many professional organizations and the recipient of numerous honors, including the George Schlesinger Prize of the city of Berlin. His research activities and publications involve manufacturing processes, equipment and systems, and the future of manufacturing technology.

IF we are to examine the technological aspects of the factory of the future, it is important that the definitions from which we are working are clearly understood. Webster defines a factory as "A building or collection of buildings with facilities (as power-driven machinery) for the manufacture of goods often from raw materials: a place where work is done in the fabricating of goods, wares or utensils." Thus the prime functional activity of a factory is manufacturing, which Webster defines thus: "To make (as raw material) into a product suitable for use: to produce according to an organized plan and with division of labor."[1]

A more specific and technologically oriented definition of manufacturing, quite consistent with Webster's, is the international one recently developed by Computer Aided Manufacturing—International (CAM-I): "A series of interrelated activities and operations involving design, materials selection, planning, production, quality assurance, management and marketing of discrete consumer and durable goods."[2]

Here the interrelated activities and operations of manufacturing that are primarily technological stand out clearly, namely such elements as design, materials selection, planning, production, and quality assurance. These are the aspects of the factory of the future this article will particularly concentrate on.

It is important to note that Webster's definition implies, through use of the words "organized plan", and CAM-I's specifically states, that the activities and operations of manufacturing are interrelated. Thus one should be able to deal with manufacturing as an integral system. This fact lies behind the major and overriding technological trend in manufacturing today, the trend toward what has been called the computer integrated automatic factory.

THE TECHNOLOGICAL TREND

In industrial countries throughout the world today there is a pronounced trend of increasing efforts in development and implementation of advanced manufacturing technology. This technology includes such new types of machining processes as laser machining, such new types of materials as composites, and such new ways of organizing manufacturing as cellular manufacturing. However, by far the most significant and important type of advanced manufacturing technology being developed and implemented today is that commonly called computer integrated manufacturing. It has already demonstrated greater potential to improve manufacturing capability than the combined potential of all other known types of advanced manufacturing technology!

What is the technological basis for this overriding importance of computer integrated manufacturing? It is threefold. First, it is the fact that the computer has unique potential to provide manufacturing with two powerful capabilities, never before available: on-line variable program—flexible—automation, and on-line moment-by-moment optimization.

Second, and very important, the computer has the capability to do that not only for the hard components of

1. *Webster's Third New International Dictionary of the English Language Unabridged* (Springfield, MA: G. C. Merriam Company, 1967).

2. *Glossary of Computer Aided Manufacturing Terms,* 2nd ed. (Arlington, TX: Computer Aided Manufacturing International, 1978).

manufacturing—the manufacturing machinery and equipment—but also for the soft components of manufacturing —the information flow, the data bases, and so on. Third, and of utmost importance, the computer has the capability to do that not only for the various bits and pieces of manufacturing activity, but also, since the computer is the world's first true systems tool, for the entire system of manufacturing. Thus it has tremendous potential to integrate that entire system, producing what may be called the computer integrated manufacturing system. This generic concept of the computer integrated manufacturing system is shown in Figure 1.

The basic importance of this concept is increasingly recognized by industry as key to meeting successfully the challenges posed by the economic and social pressures impacting manufacturing today. In fact a group of top U.S. corporate executives recently stated that "systems integration is, for many companies, 'the key to survival in the next decade.'"[3] This concept guides today's strong technological trend toward development and implementation of computer automated, optimized and integrated manufacturing. This powerful technological trend appears today to be by far the most pervasive factor that will determine the technological aspects of the factory of the future.

THE TECHNOLOGICAL FORECASTS

The virility of this trend is attested to by various technological forecasts made over the past 10 years. For example, in 1977 and 1978 the Society of Manufacturing Engineers produced three Delphi-type forecasts[4] on the future of manufacturing covering material removal, manufacturing systems, and assembly. Out of the 133 events forecast there, 99—or 74 percent—strongly forecast the coming of computer integrated factories. The selection quoted here gives an excellent feel for the timing and direction to be expected of this highly important technological development:

By 1987 (median)
Approximately 15 percent of total machine tool production will not have a "stand-alone" use, but will be a part of a versatile manufacturing system, featuring automatic part-handling between stations and controlled from a central process computer.
By 1990 (median)
The development of sensory techniques will enable robots to approximate human capability in assembly. Computer-aided design techniques will be used in the design of 50% of the new assemblies designed.
By 1995 (median)
Fifty percent of the direct labor in automobile final assembly will be replaced by programmable automation.

It can be seen that these forecasts portend the rather full realization of the computer integrated factory before the end of this century. One may well ask why this developing technology of computer integrated manufacturing is advancing so rapidly and is expected to be so rapidly implemented. As we all know, technological trends alone are not sufficient to propel industry, either in this country or abroad, into such sweeping changes as this. There have to be other

3. "Manufacturing's Sweeping Computer Revolution." *Production,* 84(6): 74-82 (Dec. 1979).

4. B. Colding, L. V. Colwell, and D. N. Smith, "Delphi Forecast of Manufacturing Systems," "Material Removal" (May 1977), and "Assembly" (May 1978), Society of Manufacturing Engineers, Dearborn, MI.

FIGURE 1
THE COMPUTER INTEGRATED MANUFACTURING SYSTEM

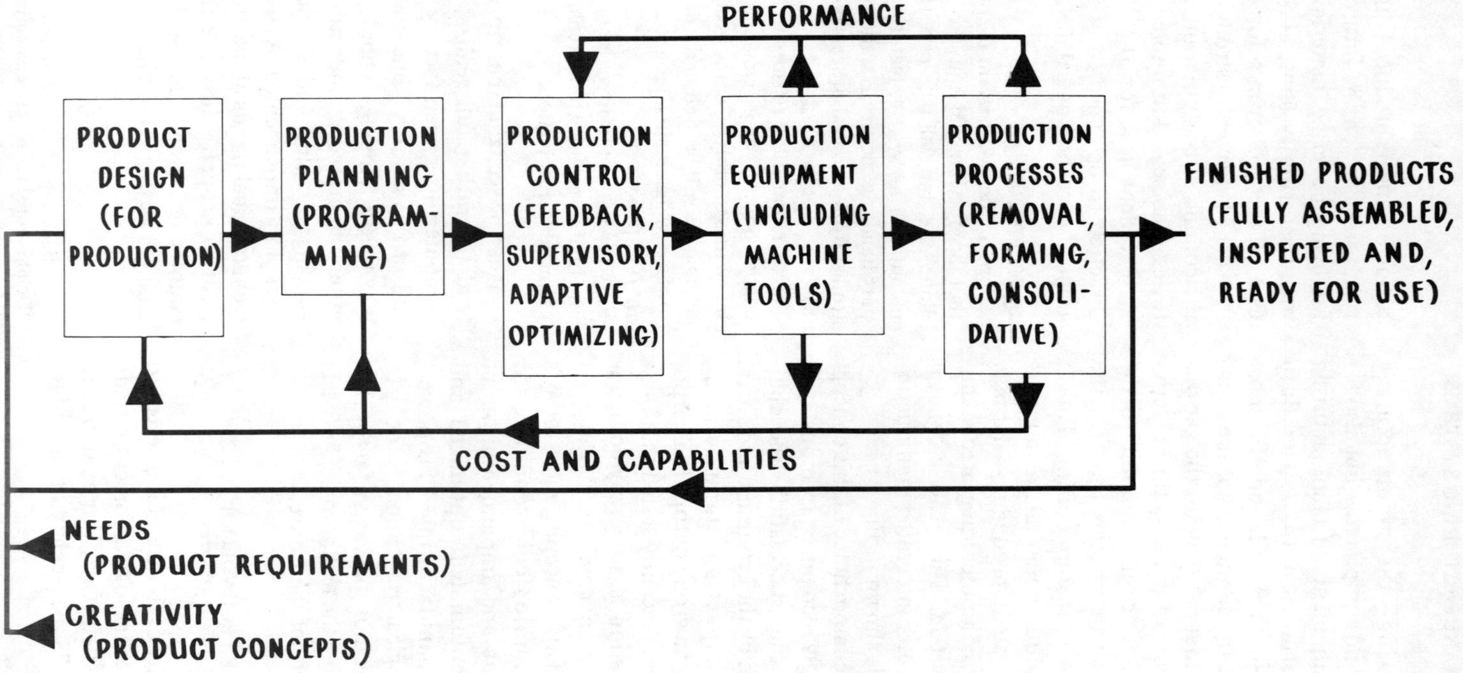

SOURCE: Cincinnati Milacron

driving forces at work—and there are. These are economic and social in nature. An example of each should suffice to illustrate their character.

ECONOMIC AND SOCIAL DRIVING FORCES

The major long-term economic force at work today in industrialized countries, as documented by the Society of Manufacturing Engineers,[5] is the continuing need to reduce the cost of creating real, tangible wealth. The reason for this fact is that manufacturing is the principal real-wealth-producing activity of such nations today. While manufacturing accounts for about one-third of the gross national product (GNP) of the typical industrialized country today, services account for about one-half. While services are essential to the support of a high standard of living and quality of life, they do not create real, tangible wealth; yet such wealth is the basis and source of all other wealth in a nation.

Thus it follows that manufacturing is responsible for about two-thirds of the basic wealth-creating activity of a typical industrialized country, the remainder of the real-wealth creation coming from the extractive—agriculture, fishing, and mining—industries and the construction industry. In the case of the United States today, manufacturing accounts for about 24 percent of the GNP, and the extractive and construction industries account for about 13 percent. On the other hand, the service industries account for about 63 percent of the GNP. However, only the 24 percent of GNP represented by manufacturing and the 13 percent represented by the extractive and construction industries result in creation of real tangible wealth. Thus, in the case of the United States, manufacturing accounts for about 65 percent of the direct real-wealth-producing activity of the country, or about two-thirds.

Since creation of real tangible wealth is the basis and source of all other wealth in a nation, increases in the standard of living, quality of life, employment, and the general economic well-being of a country stem directly from decreases in the cost of creating its real wealth. With manufacturing accounting for two-thirds of that wealth creation in industrialized countries, it follows that reduction of the cost of manufacturing is a top priority item in such countries.

How does one go about increasing the cost-effectiveness of manufacturing? Primarily by increasing manufacturing productivity. And how does one increase manufacturing productivity? Primarily through improvement of manufacturing technology. This fact has been substantiated recently in three independent studies of the relative contributions of labor quality, capital growth, and improved technology to improved productivity. These studies were conducted by Denison, by Kendrick, and by Christensen, Cummings, and Jorgenson, and their findings have been correlated and reduced to comparable terms by Morrison and McKee. The results are shown in Table 1.

It can be seen that in each case, or on average, technology was found to be the leading contributor to productivity improvements. Thus the prime contributor to increased cost-effectiveness in manufacturing—and thus to improvement of a nation's future economic and

5. "The National Role and Importance of Manufacturing Engineering and Advanced Manufacturing Technology" (Position paper by the Society of Manufacturing Engineers, Dearborn, MI, May 1978).

TABLE 1
CONTRIBUTION TO PRODUCTIVITY INCREASES (percentages)

	Labor	Capital	Technology
Denison	18	20	62
Kendrick	10	18	72
Christensen, Cummings, and Jorgenson	14	42	44
Average	14	27	59

SOURCE: D. L. Morrison and K. E. McKee, "Technology—For Improved Productivity," *Manufacturing Productivity Frontiers*, 2(6):1-8 (June 1978). Published by Manufacturing Productivity Center, Illinois Institute of Technology Center, Chicago.

social well-being—is the research, development, and implementation of innovative, advanced manufacturing technology. Of all the kinds of advanced manufacturing technology being researched, developed, and implemented today, computer integrated manufacturing has already demonstrated far greater potential to increase manufacturing productivity than anything that has appeared on the scene since the onset of the Industrial Revolution.

The major long-range social force affecting the field of manufacturing today is the trend toward postindustrial society, as described by D. Bell.[6]

This trend in turn is producing important trends in attitudes toward manufacturing. The most significant of these is the changing attitudes of workers toward conventional manufacturing jobs, namely their increasing reluctance to work in such jobs. There is a steadily increasing preference by workers entering the work force in industrialized countries for employment in the service industries instead of in manufacturing.

Bell points out that this same type of trend in work preferences occurred earlier, in the transition from the preindustrial to the industrial society. For example in the United States, as the manufacturing industry developed, the percentage of the work force employed in agriculture declined from 90 percent in 1790 to 4 percent in 1976. Meanwhile the percentage of the work force employed in manufacturing rose correspondingly during the nineteenth century. In recent years, however, it has begun to decline, from 30 percent of the work force in 1947 to 24.6 percent in 1970 and to 21.7 percent in 1978. A Rand Corporation forecast projects that by the year 2000 only 2 percent of the labor force will be employed in manufacturing; Bell's own estimate of this last figure is a more conservative 10 percent. Thus in all the industrialized countries of the world there is an increasing shortage of willing, capable manufacturing workers.

The migration of the labor force puts direct social pressure on the manufacturing industry to increase productivity at a rate more than sufficient to compensate for the shrinkage in its labor force, which is moving out of manufacturing into the seemingly more attractive service jobs at a rate expected to approach better than 3 percent per year.

6. *The Coming of Post-Industrial Society; A Venture in Social Forecasting* (New York: Basic Books, 1976).

Fortunately computer integrated manufacturing has already demonstrated major potential to mitigate this pressure. It has done so by demonstrating tremendous potential to automate nearly all aspects of manufacturing, so that highly productive manufacturing can continue with a smaller and smaller percentage of the labor force in more and more highly computer integrated factories.

As a result of such long-term economic and social forces acting on manufacturing, large national programs of research, development, and implementation of computer integrated manufacturing are presently under way in many industrialized countries. These are usually carried out by joint government-university-industry cooperation. An understanding of the nature of these can throw considerable light on the technological aspects of the factory of the future and the technological evolution that is leading toward that factory.

THE TECHNOLOGICAL EVOLUTION

Realization of full computer integration of manufacturing is the long-range goal of these national programs. It is well realized, however, that to get from today's industrial methods, know-how, and installed equipment to that goal requires an evolutionary, rather than a revolutionary, process. The strategy being followed, therefore, is to develop and implement a series of viable economic steps, in the form of shorter-range programs of research, development, and implementation, each having two main characteristics: potential for sufficient economic return to justify it by itself and to generate the capital to support development and implementation of the next step; and compatibility with the eventual attainment of the goal of full computer automation, optimization, and integration of manufacturing.

Out of the variety of programs being pursued, the following seem to be receiving a major part of the attention and effort worldwide:

1. Development of integrated manufacturing software systems, through development of individual software modules that can in the long run be readily interfaced with each other to build up full software systems appropriate to various types of manufacturing applications. Such work is proceeding rapidly in the Council for Mutual Economic Assistance (COMECON) countries, as well as in West Germany, Norway, and to a lesser extent Japan. The U.S. Air Force is carrying out a major program of this type, its Integrated Computer Aided Manufacturing (ICAM) program. CAM-I, with headquarters in Arlington, Texas, has under way several programs to develop integrated manufacturing software systems.

2. Development and application of group technology and cellular manufacturing as a required base for application of hierarchical computer systems and the evolution of flexible manufacturing systems. Group technology involves analysis of the various characteristics of manufactured parts for the purpose of identifying common characteristics, which allow them to be grouped into families. These families can then be manufactured in cells, equipped to deal with the minor variations in part characteristics, almost as though all the parts were identical. Such work is proceeding most rapidly in the Netherlands, Japan, West Germany, Norway, Britain, and the COMECON countries. The U.S. Air Force ICAM program also has this as one of its features, as does the CAM-I program.

3. Development and application of broad computer control of machine tools and manufacturing processes, through use of numerical control (NC), computer numerical control direct numerical control (DNC), (CNC), and hierarchical computer systems. Such work is proceeding most rapidly in Japan, the United States, West Germany, and Norway.

4. Development and application of computer-controlled robots for automating workpiece and tool handling, machine operation, and manufacturing processing of various kinds. Such work is proceeding most rapidly in Japan, the United States, West Germany, Norway, Sweden, the USSR, Bulgaria, and Hungary.

5. Development and application of computer-controlled flexible manufacturing systems that are in effect rather fully automated group technology cells. They consist of groups of machines served by automated tool and workpiece transport and handling equipment, all operating under computer control to produce rather broad families of parts. Such work is proceeding most rapidly in Japan, the United States, and East Germany, and more recently in West Germany, the USSR, Bulgaria, Czechoslovakia, Hungary, Norway, and Britain.

6. Planning and development of prototype computer automated factories themselves. Such work is proceeding most rapidly in Japan, which has undertaken a major national program to develop a prototype minifactory by 1984. In addition, several West European and COMECON nations, including the USSR, have general plans and supporting programs aimed specifically at this objective.

To give some feel for the overall nature of the work being done in such programs, I will describe two of the specific programs being carried out by two quite different countries—Norway, a relatively small country, and Japan, a major world power.

Norway's primary national program has strong economic, social, and defense implications and considerations. To understand these it is necessary to understand certain current economic and social conditions in Norway. The Norwegian fishing industry is deteriorating rapidly. Further its agriculture is becoming rapidly mechanized. These factors are leading to severe unemployment in the rural areas of Norway. Meanwhile in its cities there is a growing shortage of workers willing to work in its still relatively small manufacturing industry. This is putting tremendous pressure on the rural population to move to the cities, which is the last thing Noway wants. Not only would the economic and social costs of developing huge metropolitan complexes be great, but it would also leave the rural areas relatively defenseless. Thus the Norwegian government has requested proposals for programs to solve these problems.

One of the most promising of these proposals, so far, is one put forward by Ø. Bjørke[7] of the Technical University of Norway (NTH-SINTEF) in Trondheim. It is aimed at developing a strong and considerably expanded manufacturing industry in Norway, but one located primarily in the rural areas. The prime concept making this location possible is an extension of group technology.

Each part-family cell in a manufacturing plant laid out on group technology principles—as many Norwegian fac-

7. "Computer Aided Part Manufacturing," *Computers in Industry*, 1:3-9 (1979).

tories already are—is in effect a minifactory within a factory. Bjørke's concept is to decentralize manufacturing plants by dispersing their individual group technology cells for parts manufacture into the rural areas, as a network of minifactories, providing employment there.

However, Norwegian wages are now almost the highest in the world. Therefore, in order for an expanded Norwegian manufacturing industry to be competitive in, and capture a significant share of, the world market for manufactured goods, its productivity and efficiency must be very high and its manufacturing costs very low.

Thus the second concept in Bjørke's plan is to provide each minifactory with a high-technology core, one that runs 24 hours a day producing parts automatically. Each such core would be supported, during one 8-hour shift per day only, by a work crew of 5 to 15 or more persons. These would do such work as computer assisted programming, production planning and scheduling, heat treatment, off-line processing not readily done automatically in the cell, mounting and demounting parts on pallets, and so on. The core would run unattended during the other two shifts; it is already illegal in Norway for anyone to work a third shift in the manufacturing industry. Figure 2 shows Bjørke's cell concept.

The high-technology core that Bjørke proposes consists of a group of four or five NC machine tools and machining centers, suited to production of the particular part family, arranged in a circle around a single robot, which does all the part handling and machine loading and unloading in the cell. It is a DNC cell, with CNC of the machines and robot. Figure 3 presents a schematic of the prototype high-technology cell core currently operating in his laboratory.

Now let us turn our attention to Japan. Their programs tend to have strong emphasis on manufacturing hardware. As a result Japan already has accomplished major implementation of computer-controlled flexible manufacturing systems in industry, and some of these are the most advanced in the world, running unattended at night. This is due in large part to a very favorable government climate for development and implementation of such advanced manufacturing technology — a climate that includes considerable government assistance, both indirect and, to some extent, direct.

One of the most ambitious of the government-supported programs in this field has been that known as Methodology for Unmanned Manufacturing (MUM). This was begun in 1972 and was carried on cooperatively by a consortium of government agencies, technical societies, universities, trade associations, and industrial companies. It produced a proposal and plan for developing and building, with $100 million of government funding, a prototype so-called unmanned factory—a work crew of about 10 persons instead of the usual 700 to 800—for production of machine parts and components. The proposal and plan were completed at the end of 1975 and submitted to the Ministry for International Trade and Industry (MITI) for funding. It failed to receive funding because it was considered too idealized a plan and because of political pressures from two other groups, one pushing for a major program on laser machining and one for a major program on work preparation in manufacturing.

The result was a new program initiated in 1977, combining aspects of the laser machining and work preparation programs with that for the unmanned

FIGURE 2
THE NORWEGIAN INSTITUTE OF TECHNOLOGY CELL CONCEPT

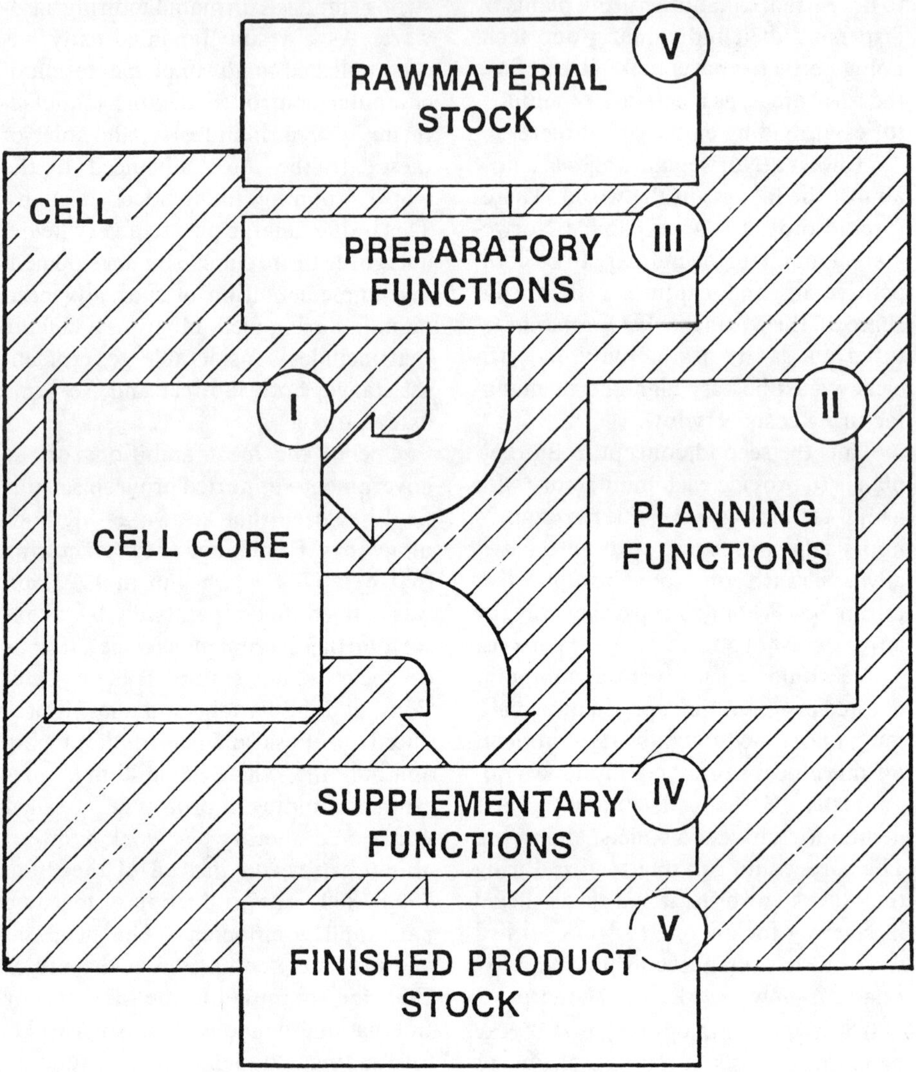

SOURCE: Ø. Bjørke, "Computer Aided Part Manufacturing," *Computers in Industry,* 1:3-9 (1979). Reprinted by permission of author.

factory.[8] The former MUM plan is now considered to be the idealized plan that will provide general guidelines for work in this area for about the next 20 years, while the new program is considered to be the practical plan. Three government laboratories and 20 manufacturing companies are participating in it, and this work is being guided by the Agency of

8. "Research and Development of an Ultra-High-Performance Laser-Applied Complex Production System," *Technocrat,* 10(12):13-21 (Dec. 1977).

FIGURE 3
THE NORWEGIAN INSTITUTE OF TECHNOLOGY CELL CORE

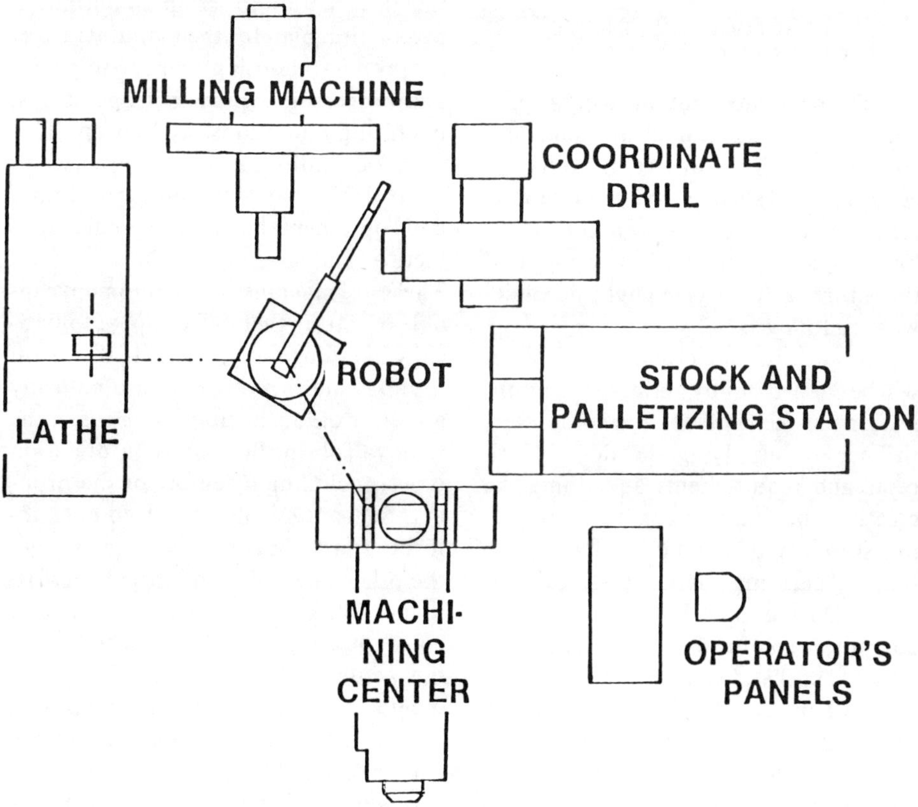

SOURCE: Ø. Bjørke, "Computer Aided Part Manufacturing," *Computers in Industry,* 1:3-9 (1979) Reprinted by permission of author.

Industrial Science and Technology of the Ministry of International Trade and Industry.

The program is targeted primarily at reduction to practice of existing technology rather than having the heavy dependence on the development of new technology featured in the MUM plan. Further, one of its guiding principles is to design the hardware so that there will be minimal need for new software. As such it has almost no new software development in it. The program is still expected to culminate in a small prototype rather fully computer automated model factory about 750 square meters in size, by 1984, the year of termination of the project. It would employ about three or four operating workers, plus a somewhat larger support crew than that envisioned for the MUM unmanned factory. The program has been granted about $50 million funding by MITI for the seven years. Commercialization of results will begin in the mid-1980s. The current status of the program was most recently described by M. Kimura and his associates.[9]

TECHNOLOGICAL VIGNETTE OF THE FUTURE FACTORY

Taking into account the insights provided by the technological trend and forecasts, the long-term driving economic and social forces, and the current technological evolution, it is possible to create a technological vignette of the future factory. This might be somewhat as follows.

The process of designing a product will be carried out by iterative communication between persons and computers, the person supplying the design concepts and requirements and doing the creative work, and the computer supplying stored and standardized information and carrying out the design calculations. During this design process the computer can, through the internal loop labeled "cost and capabilities" shown in Figure 1, be constantly calling for and taking into account information on the manufacturing costs and capabilities—equipment and process costs and capabilities—required to produce, on the true production elements of the available system, each of the alternative features conceived by the designer. The computer then uses that information to find a design that not only satisfies the product requirements but can also be manufactured at minimum cost and with maximum producibility.

9. "A Flexible Manufacturing System Complex Provided with Laser (FMC)—A National R & D Program of Japan" (Paper presented at the Twenty-third International Machine Tool Design and Research Conference, University of Manchester Institute of Science and Technology, Manchester, England, 14-15 Sept. 1982); the Proceedings of the conference to be published by Macmillan Limited, London, in 1983.

Almost simultaneously the production planning part of the system is using this information to set up an optimized production plan for the manufacture of the product, choosing proper equipment and processes, sequence of operations, operating conditions, and so on. This numerical information is in turn used to control the array of automatic machines and equipment that do the actual manufacture and assembly of the product. These are machines and equipment capable of setting themselves up, automatically handling parts, selecting their own tooling, and carrying out automatically a variety of fabrication processes of the removal, plastic flow, and consolidation types, including assembly of the product. Self-optimizing, they feed back information to the control system through the other internal loop labeled "performance" in Figure 1. This system, as it constantly receives this information of the actual performance of the equipment and processes, compares this with the ideal performance planned in the earlier phase. Then, as it finds that performance beginning to depart from the planned optimum, it overrides the original plan, performing dynamic scheduling, adjusting operating conditions of the machines and processes and so on as necessary to maintain optimum, minimum-cost performance.

Meanwhile the machines and equipment carry on self-diagnosis of their condition, and where an impending malfunction is detected they carry out appropriate corrective action, including automatic replacement of defective modules in the system. Further the machines also carry out automatic in-process inspection of the product at each stage of its manufacture, so that any impending deviations from the original specifications are automatically corrected and held within prescribed tolerances. Thus the final assembled product is turned out fully inspected and in full conformity with the original design concepts and requirements.

THE CHALLENGE TO AMERICAN INDUSTRY

Enticing prospects for extraordinarily enchanced manufacturing productivity are intimated by this vignette. Furthermore major efforts in research, development, and implementation of computer integrated manufacturing are already under way worldwide. These facts pose at least two major challenges to American manufacturing industry.

First, the lion's share of the basic knowledge and innovative technology pertinent to computer integrated manufacturing has been developed in this country. Yet, sadly, American manufacturing industry has not moved as rapidly as industry in some other countries, most notably Japan, to implement the results of that native basic knowledge and innovative technology. In fact such basic knowledge and technology have all too frequently been spurned by American manufacturing industry while being eagerly acquired and skillfully applied by industry in other countries. American manufacturing industry can continue to be contemptuous of such fundamental, highly innovative manufacturing knowledge and technology, and the opportunity for major reindustrialization which it presents, only at utmost peril to itself.

Second, a major factor behind the speed and skill with which industrial concerns in most other countries acquire and implement capability in computer integrated manufacturing is the fact that they are willing to draw fully on the major national resources pertinent to the pursuit of such efforts, namely universities and government. As a result a close working partnership between industry, universities, and government, in pursuit of the goal of full realization of computer integrated manufacturing in industry, has grown up in numerous countries. In particular the capability of universities to contribute effectively in this effort has been found to be a prime factor in achieving rapid progress toward realization of the goal. Here is a major resource that most U.S. companies have failed to take advantage of, or even recognize, much to their detriment. It is of utmost importance that American industry take a thorough look at the potential of that major resource, as it moves toward the pursuit of enhanced manufacturing productivity through computer integrated manufacturing, and carefully plan a strategy to take advantage of it. Efforts to enchance manufacturing engineering education and research in American universities needs to be a major part of that strategy.

The day about to dawn threatens to be one of overseas superiority in manufacturing productivity. Will we let it be so?

ANNALS, *AAPSS*, **470**, November 1983

The Unmanned Factory and the Community

By LOUIS A. FERMAN

ABSTRACT: The impact of unmanned factories on local communities must be viewed from two perspectives: (1) the historical trends that have disengaged the factory, smokestack or unmanned, from the local community, and (2) the influence of the new technology on community organizations. From a beginning of close association, the factory had disengaged from the local community through the influence of absentee ownership, technology, corporate policy, and managerial mobility. The new technology of the unmanned factory augments these tendencies toward disengagement. We have only speculative information on how the unmanned factory affects the local community, but three broad categories of direct and indirect impacts are indicated: (1) jobs and labor force, (2) the web of suppliers and other supportive firms in the community, and (3) the psychological and medical health of the community.

Louis A. Ferman, Ph.D., Ann Arbor, Michigan, is professor of social work and research director, Institute of Labor and Industrial Relations, University of Michigan. He is a consultant in employee involvement programs in industry and outplacement programming, and editor of Policy Papers in Human Resources and Industrial Relations. *Among his books are* Poverty in America, *with J. Kornbluh and A. Haber;* Mental Health and Economy, *with J. Gordus;* Economic Failure, Alienation and Extremism, *with M. Aiken and H. Sheppard; and* Plant Closing and Economic Dislocation, *with J. Gordus and P. Jarley.*

THE purpose of this article is to explore the relationship between a new form of factory organization, technology intensive rather than labor intensive, and its local community environment. The image of the unmanned factory has been set forth in popular and scientific literature: banks and banks of machines, some humming and buzzing, others silent, each performing a continuous set of operations that have been programmed into the machines at some earlier time. What is even more important, there are only a few workers who watch over this mechanical complex—"button pushers" and "button fixers."[1] Although there have been frequent inquiries into the impact of this form of factory organization on the individual worker, on the corporation, on middle management, and on the unions, the literature on how the unmanned factory will affect the local community is almost nonexistent. There is some speculative agreement that it will act to reduce jobs and tax revenue for the local community, but nothing more has been said.

The reasons for the neglect of this topic are easy to discern. First, the workers, the corporation, and the unions are the first to feel the impact of the unmanned factory; thus research has focused on them, the actors in the immediate firing line. Second, there is no community as yet that has been inundated with workerless factories, so that the phenomenon of a single unmanned factory buried in a forest of smokestack factories is apt to mask or hide the impact on the local community. Finally, the advent of the unmanned factory as an affordable and realizable form of organizational technology is still for the most part in the future, making it difficult to isolate effects from empirical observations.

An important consideration in exploring the relationship between the unmanned factory and the community is to understand the historical context in which factory-community relationships have evolved in the United States. Steeped in early paternalism and close association, the relationship evolved to one of recent disengagement and distance even before the advent of robotics and high technology. In one sense the impact of the unmanned factory on the community may be less a factor of its technology and more a consequence of historical trends in factory-community relationships. Moreover this new form of technological organization, particularly if it becomes widespread, can have a number of consequences for the local community beyond the loss of jobs and taxes. It can, for example, influence the kind and quality of leadership that factories provide to the community. It also can place demands on the local educational systems for educational and training facilities beyond those required by smokestack factories. And of course it can influence the industrial mix of a community.

What seems clear is that there are two lines of influence that must be noted: (1) historical traditions of factory-community relations, and (2) the impact of this particular type of technology on the community.

1. D. McLaughlin, "Technology and Jobs," in *Contemporary Studies in Energy Analysis and Policy,* ed. Gladys Gruenberg (Greenwich, CT: JAI Press, 1983).

THE HISTORICAL CONTEXT OF FACTORY-COMMUNITY RELATIONS

The factory-community relationship has evolved over time. The early New England mill factory, epitomized by the "Lowell girls" in Lowell, Massachusetts, in the early 1800s, was tied to its town by religious, moral, and economic bonds. There was a close identity of interests, and one was the extension of the other. The factory girls were drawn from the rural hinterland, lived in boarding houses supervised by the local minister and owned by the mill owner, and literally had their welfare entrusted to people associated with the factory. Many of these young women had their earnings saved for them to build a dowry for a marriage. In this sense the factory was both a moral and an economic entity, and was reinforced by religious and ethnic bonds.

The horror of the factory system in New England, as we have come to know it, began around the 1830s, from an increased worldwide demand for textiles and the filling of new labor demands by French Canadian women. The bond of common religious and ethnic identity between the owner and worker was broken, with a consequent increase in the owner's exploitation of workers. Children were used extensively to fill out the work force, a practice that grew after the Civil War into the early years of the twentieth century, until a flurry of child labor laws greatly reduced it.

In the post-Civil War period, economic and social forces were at work to reduce the early bonds of social intimacy between factory and community. In their classic study of the shoe factories in Yankee City, W. L. Warner and J. O. Low note the shifts in the relationship.[2] In the early and mid-1800s the shoe factories were owned by local entrepreneurs in the community, most of whom worked side by side with their employees. The bonds between community and factory were close, and it was not unusual for the factory owners or their kin to dominate the social and political life of the community. Indeed the owners of the shoe factories frequently held the major political offices of the community. The class system of Yankee City reflected the skill hierarchy within the factory: a person was known socially by the work he or she did in the factory. Certain skills were handed down intergenerationally from parent to child, and such families were frequently identified socially by the possession of such skills.

In the latter half of the nineteenth century there were two sharp changes in factory-community relationships in Yankee City. The need for capital to expand the factory brought in a new class of ownership—the absentee owner—and a new style of management—the appointed plant manager who worked for a salary and did not own the factory. Frequently the plant manager came from outside the community and viewed the community as a place of residence rather than a place of origins. The second change was technological: machinery was brought in to produce shoes formerly made by hand. For Yankee City these changes meant (1) the loss of personal contact between owner and worker with common resi-

2. *The Social System of a Modern Factory* (New Haven, CT: Yale University Press, 1947).

dence in the community, and (2) the literal destruction of the community status system that had previously depended on the factory skill system. Key decisions on factory operations were no longer made in the community but by the "Wall Street bankers" or the "New York crowd." In a similar vein, job assignments were no longer based on skills resident in local families but on the demands of the new technology. Similar patterns in other communities at the time resulted in a growing disengagement and distance between the factory and its community environment.

The restructuring of factory-community relationships by absentee ownership and its consequences has been noted by R. Stern and H. Aldrich:

> The creation of absentee owned corporations created a calculus of corporate welfare separate from that of community welfare. Though the corporation still required the cooperation of local governments and labor forces, the community simply became a geographic location in which to operate as long as relative cost advantages were maintained. Commitment to the community itself and knowledge of the local social structure declined because professional managers sent to run these branch plants were more concerned with their careers in the larger corporation than with the quality of community life.[3]

R. O. Schulze has noted further consequences of the absentee ownership phenomenon in his study of Ypsilanti, Michigan:

> As the activities of these units became increasingly directed toward—and by—populations and groups other than the local ones, the relevance of local community organizations and the impact of local political influences on the major economic units would accordingly diminish. As this occurred, the local power structure would in effect bifurcate—with those who exercised primary direction over its socio-political system no longer being essentially the same set of persons who exercised primary control over its economic system.[4]

Although local owners began to disappear, the new managerial class resided in the community and participated in community life, but their concerns were different. These managers became a new thread between the factory and community. Schulze went on to note that this engagement of managers in community life was supported by corporate policy. The good manager was one who joined local organizations and lent his organizational talents and resources to the solutions of problems within the community. The corporation viewed such involvements as good public relations and as protective of corporate interests.

In many cases these managers together with the local professionals monopolized the political life of communities and held the major political offices. New forces were to reshape the behavior of the managers. As communities like Ypsilanti increased in ethnic and social diversity, there was no longer a single political grouping; rather there developed a nucleated political system with many diverse political groupings mobilized around controversial issues. Corporate policy shifted in regard to management involvement in local community affairs.

3. "The Effects of Absentee Firm Control on Local Community Welfare: A Survey" (Paper presented at the FTC Bureau of Economics Conference on Firm Size, Market Power and Social Performance, Washington, DC, 17-18 Jan. 1980).

4. "The Role of Economic Dominants in Community Power Structure," *American Sociological Review,* 23:3-9 (June 1958).

Managers were instructed to play it safe and not get involved in controversial issues or elective office, lest the corporation's interests be compromised by shifts in political currents. Managers could still be involved, but in noncontroversial issues—United Way campaigns or Red Cross drives.

Another current was the decentralization of business offices, with locations in a number of communities. This spatial diversification meant that the company paid less attention to the social and political situation in any one community, indeed reducing its overall commitments to the community locations of its business. The end result was the weakening of ties between factory and community, and the managerial abdication and relegation of political power to the professional elites whose interests and views did not exactly coincide with those of plant managers. For example in many cases the professionals opposed zoning and environmental positions that might have been favored by the managers.

J. M. Sheppard and J. G. Houghland[5] have advanced the thesis that new patterns of managerial mobility have also acted to erode close ties between factory and community. The early class of managers, who replaced local owners, put down roots in the local community and frequently had a long, if not a lifetime, period of residency. This resulted in commitments. The current class of managers has a shorter period of residency and is oriented to a career and the many outposts of the company. Consequently the commitment to the local community is shallow and temporary.

Other factors also reinforced this historical disengagement of the factory from the community. The first of these was the life cycle of a factory. W. Thompson has argued from locational evidence that as a factory ages it goes through a number of evolutionary stages.[6] In the early stages the factory is rooted to a given location, with close developmental ties to the local community. At a much later stage the factory becomes less dependent on local resources and can be moved anywhere, introducing a source of potential instability in the economic life of the community.

Thompson points to early locations of factories close to urban centers, locations chosen for the sake of the critical urban core of expertise and support services. At a later period, when new transportation and communication linkages were developed, the factories were no longer rooted to the old location and dependent on its resources but were given latitude to move, usually to suburban areas. In the present period still further improvements in technology, communication, and transportation have made it possible for a factory to move to the rural hinterland where land is cheaper, where there is room for expansion, and where dependable, low-cost labor is available. This life cycle of a factory, then, suggests that spatial mobility rather than immobility will be the rule, and that there will be destabilizing pressures on any factory-community relationship as the factory ages.

5. "Organizational Size, Managerial Mobility and Corporate Policy: A Study of the Community Participation of Managers," in *The Impact of the Modern Corporation: Conference Papers* (New York: Columbia University Center for Law and Economic Studies, 1982).

6. "Industrial Location: Causes and Consequences," in *Michigan's Fiscal and Economic Structure,* ed. H. E. Brazer (Ann Arbor: University of Michigan Press, 1982).

Two other sets of factors have increased the spatial mobility of factories and weakened their ties to local communities. The first set is locational. Old, established factories have relocated for a number of reasons: (1) to move closer to a market for their products, (2) to avail themselves of a cheaper source of labor, (3) to move to a source of raw materials, and (4) to take advantage of tax or rental arrangements offered by competing communities. In the last 50 years there have been great shifts in industrial activity—textile, automotive, steel—from the northeast and midwest to the southeast and west. These locational shifts have left a trail of industrially denuded communities with large numbers of unwanted workers.

The second set of factors is technological. New technology has made the factory more mobile spatially. The computer chip is the latest in a long line of technological advances that has reduced the bulk of factory machinery, making it possible to relocate operations from one community to another with little cost or effort. Advances in communications permit multiplant companies to shift production work at the touch of a button from one facility to another many miles away. The result was the introduction of community uncertainty about the stability of local employment levels and productive activity.

These historical developments in ownership, corporate policies, managerial mobility, and technology weakened the ties between the factory and the local community. This estrangement had implications for the stability of the local industrial base, opportunities for local workers, and the input of financial resources into the community. These historial trends are strong and affect smokestack as well as unmanned factories. Whatever role the new technology plays in relations between factory and community, the American historical experience and its pressures on the current situation cannot be ignored.

THE IMPACT OF THE TECHNOLOGY-INTENSIVE FACTORY ON THE COMMUNITY

Technological change in industry can have a profound effect on a local community. We have already discussed how the mechanization of shoemaking in Yankee City had the effect of destroying the family-based skill hierarchies of the community.[7] Research in the 1940s documented the erosion of whistle-stop communities when trains switched to oil and no longer needed periodic community stops to supply water and other services for the coal burners—literally a "death by dieselization."[8] The dying ore communities of the Upper Peninsula of Michigan that followed the exhaustion of the mines by technology-intensive drilling operations are other examples.

There is no empirical research as yet on how the unmanned factory affects its community. An initial and still controversial question is the extent to which such factories create or destroy jobs—an issue of considerable concern to the local community. Depending on the assumptions made about growth in high technology and robotics, there are almost as many opinions on the question as there are researchers.

On the front page of a recent *New York Times* there are back-to-back articles reporting on the deliberations of two

7. Warner and Low, *Social System of a Modern Factory*.

8. W. F. Cottrell, "Death by Dieselization," *American Sociological Review*, 16:358-65 (June 1951).

conferences—one of manufacturers who were potential purchasers of high technology and robotics equipment, and the other of producers of high technology and robots.[9] The first group saw a marked decline in the need for labor when such equipment was introduced, while the second saw a need for different workers but no reduction in the need for labor. Another *New York Times* article reviews the contradictory views on the impact of new technology on the job market.[10] The overall conclusion from a large number of studies is that there will indeed be employment growth, but it will be in engineering and related technical fields, intensifying the growing jobless rates in traditional white collar and blue collar areas. A recent study by the W. E. Upjohn Institute for Employment Research[11] focused on the job displacement potential of robots in Michigan. The study suggests that modest displacement will occur in industry but will be concentrated in blue collar occupations, such as painting and welding. A second conclusion is that normal attrition from factories will contain the displacement and make mass layoffs unnecessary.

These remarks suggest that the advent of the unmanned factory means a shift of skill mix—more conceptual than manual skills—some necessary displacement of workers from unskilled jobs, and a growth in engineering and technical employment. The consequences of these changes for the community may be far reaching.

First, unwanted workers require community support in terms of income maintenance, resources for training and retraining, and programs to heal the psychic wounds of displacement. The community is forced to expend effort and money to meet needs that did not exist before the advent of the unmanned factory.

Second, the community becomes a home for strangers as increasingly the factory labor, highly skilled and trained, is imported from outside the community. Besides bringing new demands for social amenities and services into the community, these newcomers pose problems of social integration.

A third consideration is the impact that unmanned factories have on small supplier firms that were established to supply parts or services to old smokestack factories. To survive, these firms must undergo changes in their own labor force and technology to service the needs of the unmanned factory. Their survival may indeed be threatened if such needs can already be met by supplier units that are in the factory or affiliated with the parent company. At any rate the investment may be beyond the resources of these old supplier firms and may lead to shutdowns, further increasing the number of displaced workers in the community.

We can divide the impacts of the unmanned factory into three broad categories:

—direct impact on the worker and labor market adjustment problems;

—wider community economic and social impacts; and

—psychological and medical impacts on the worker and community.

9. *New York Times,* 16 May 1983, p. 1.

10. "Impact on Job Market: Divided Predictions," *New York Times,* 28 Mar. 1982, sec. 2, pp. 39-40.

11. H. A. Hunt and T. L. Hunt, *Robotics: Human Resource Implications for Michigan* (Kalamazoo, MI: W. E. Upjohn Institute for Employment Research, 1982).

Impact on the worker and labor market adjustments

In the transition from the smokestack factory—with its production line and labor-intensive operations—to the unmanned factory, we can anticipate a number of effects on the community labor force.

First, there will undoubtedly be surplus workers whose work operations are no longer part of the factory. These displaced workers will be overrepresented among the unskilled and undereducated who had minimal skills and training on the production line and could not make the transition to work in the unmanned factory. There will be a need for fewer, but highly trained, workers, but most of those displaced will not have the education or background to compete for the jobs or for training opportunities. Workers who make the transition will be those who were already in sophisticated maintenance work in the older factory. We know that displaced workers have greater difficulties in reemployment, particularly if they are older, unskilled, and undereducated. About one-half of reemployed displaced workers take a job that is lower in pay, status, and fringe benefits than the last job.[12] Other displaced workers will take early retirement, engage in off-the-books work, or become marginal to the labor force as discouraged workers. What this means is that the community will be faced with the prospect of a labor force with more unemployed workers, more workers who have skidded from better jobs, and more workers who have retired into nonproductive roles.

The unmanned factory will affect not only those workers who are already in the labor force but also the youth who are about to enter the labor force. For the high school dropout as well as for many high school graduates, the entry-level jobs on the production line of the smokestack factory were the portals into the world of work. These jobs demanded little skill or education and could be learned in the workplace. In the unmanned factory such jobs will almost be nonexistent. Jobs will require formal training and at least a high school or technical education, with some emphasis on mathematics. For disadvantaged youth the unmanned factory will offer almost no opportunity structure. For the community this means a growing pool of youth, poorly equipped for the labor market, with no viable opportunity structure except in low-paying service jobs. Effort, services, and money will have to be expended to handle a problem that had previously been minimized by the entry-level jobs in the smokestack factory.

We foresee that minority groups, especially Hispanics and blacks, might be the most negatively affected by the coming of the unmanned factory, because these groups are overrepresented in the blue collar, low-skilled jobs that will be either eliminated or reduced by the new technology. Women will certainly feel the effect of a restricted opportunity structure, because the high-level jobs in the unmanned factory will require a background in mathematics and science, subjects that have not been central to curricula for women in secondary and postsecondary education. It is possible that these deficiencies will relegate women to the lower-level jobs of the factory, with higher-level jobs going to men. Many of the gains made in the

12. J. Gordus, P. Jarley, and L. Ferman, *Plant Closings and Economic Dislocation* (Kalamazoo, MI: W. E. Upjohn Institute for Employment Research, 1981).

1960s and 1970s by these groups may very well be wiped out in the 1980s.

Success in the labor market will require training as never before. The American industrial system has minimized formal classroom instruction for most jobs, emphasizing instead on-the-job training. The new mode and burden will be to establish training facilities in the community to serve the interests both of workers and of the factory in the new technology. This may involve a massive restructuring of all levels of existing community education and even the development of new structures of adult education. Training for work, which in the past had minimal significance, will assume central importance in the community of the unmanned factory.

It is widely assumed that the unmanned factory brings with it more skilled jobs of a higher occupational status. This may be true for many jobs, but there is the possibility that for some workers the new jobs may be deskilled and of a lower occupational status. The worker in the smokestack factory may have needed to carry out eight or nine manual measurement operations that required great skill. Has this job increased or decreased in skill and status if these measurements are now electronically made, and the worker simply reads and records these figures? Since skill and occupational status are the determinants of standing in the community, these questions might be of crucial importance to workers.

The unmanned factory can, then, affect the worker and the labor force in several ways. It may reduce the total number of jobs, change the content of tasks, require more formal training, change the skill or status level of the job, and reduce the number of entry-level jobs for newcomers to the labor market. In all of these the community has a central concern, because it must contend with increased unemployment, new services for displaced workers, and new structures for training and education. This concern comes at a time when local, state, and national financial support is on the decline.

Wider community economic and social impacts

Income losses to the community can be severe if the unmanned factory results in considerable job displacement. The most immediate effect is a loss of factory payroll wages; indirect losses accrue to other businesses, however, as purchases of inputs by the factory and goods and services by the employees are lost. Decreased community income can also precipitate loss of employment in other local businesses. Where unemployment from job displacement is prolonged, demands on public services will rise while tax receipts from purchases and lost businesses decline.

In any community there is a web of interdependent businesses. Some businesses supply parts for other firms, and other businesses supply services. This web would probably be broken by the advent of the unmanned factory. These supplier firms would require a change in products or services and certainly a change in technology and labor. Some would close while others would require a different labor supply from the community. A likely consequence would be more job displacement as these firms struggle to accommodate a new industrial reality.

The unmanned factory would also rely on a range of job specialties different from that of the smokestack factory. There would be less need for persons skilled in labor relations, community relations, or personnel administration, and more need for engineers and technicians. This shift may affect the community in that the factory has traditionally made available personnel who might contribute to solving problems in the community. The smokestack factory had a diversity of personnel specialties that could be used in a variety of problem situations. This diversity is reduced in the unmanned factory, and consequently there is the likelihood of a reduction in factory contributions to community problem solving. The nature of the technology makes the factory less dependent on community resources, such as labor, and might further exacerbate historical trends in disengagement of the factory from the community.

Psychological and medical impacts on the worker and community

Another important cost to the community of the unmanned factory, assuming that it does result in extensive job displacement, is the severe emotional stress created by a sudden loss of employment. The psychological stress created by the feeling of being unwanted, dislocated, or useless manifests itself in a number of illnesses, both mental and physical. Loss of income can also affect the medical health of workers and their families. Such medical problems may range from hypertension and ulcers to chronic insomnia and depression. The costs associated with these problems accrue to the individuals suffering from them as well as to their families and the larger community. They are often referred to as the noneconomic costs of job displacement, although clearly the psychosocial and medical problems have both immediate and long-term economic costs.

There are costs for both the individual and the community. For the individual there is the reduced energy for seeking work or for holding a job once it is found. This interplay between health and labor market activity can result in chaotic work careers and the onset of unemployment careers, in which joblessness becomes a continuous fact of life. For the community there is the cost of treating patients for the medical problems resulting from job displacement. Often the number of people requesting services increases dramatically. Sometimes new programs have to be established. These expenses cannot be passed directly to the jobless worker; the community at large must absorb the cost.

SUMMING UP

The impact of the unmanned factory on its community cannot be viewed only in the light of new technological systems. There are strong historical currents in factory-community relationships that suggest a progressive disengagement of the factory from community life. The characteristics of the new technology will apparently exacerbate this historical process by making the factory more mobile spatially, reducing the number of factory personnel inputs into the community, and becoming less dependent on technical and human resources from its immediate community environment. The most immediate impact on the community will be a loss of job opportunities, not only for many current workers but also for the young entrants to the labor market. This apocalypse is not around the corner, but years away. Certainly there is yet time for planning to make the local community a more organic part of the operations of the unmanned factory.

ANNALS, *AAPSS*, **470**, November 1983

Workers, Unions, and Industrial Robotics

By THOMAS L. WEEKLEY

ABSTRACT: The concerns of organized labor pertaining to technological changes have been vastly misunderstood. Labor organizations are well aware of the need for new technology and the reactions of their members. This article is offered in the hope that users and manufacturers of automated equipment may gain first-hand insight into the attitudes and concerns of organized labor and the worker affected by advanced technology. Specific worker attitudes are presented, along with various concerns of workers and labor organizations. Suggestions are offered, along with information outlining current programs designed to cope with the problem of an ever-increasing demand for goods, an ever-increasing population and work force supplied by a decreasing number of job openings.

Thomas L. Weekley has had first-hand experience in the use and construction of automated equipment, as an operator and skilled craftsman. He attended Youngstown State University School of Engineering, is a graduate of Baptist Christian College (B.A.) and has worked extensively within the United Automobile, Aerospace and Agricultural Implement Workers Union of America at local and national levels. He is currently an international representative and coordinator of the Skilled Trades Department of the UAW, and his primary duty is to keep abreast of technological change and its effects on workers.

THE United Auto Workers' (UAW) Union began in the 1930s. When did this union first address new technology? Is new technology a new question? Obviously it is not. This issue was first addressed in the 1930s, and in 1935 the United Auto Workers' first convention talked about many of the issues that are still being talked about today. The terms used were "standardized wages," "economic security," "company profit," and so on, terms still familiar today. We are facing the same problems. They are more complex, but they are still the same problems. Management's striving for profits has always conflicted with the union's drive for protecting working men and women, not only their benefits, but their security on the job. In the 1950s this really became more of a problem, and new technology and technological advances began to be especially addressed.

HISTORICAL BEGINNINGS

In a talk before a joint session of Congress on 17 October 1955, Walter Reuther gave testimony that outlined not only the union's position but some of his predictions for the future, predictions that proved to be very accurate. What he predicted in the 1950s and his reasoning have since become the basis for the UAW's position on new technology. At that time, a type of industrial revolution had begun, bringing a changeover to fixed automated pieces of equipment, such as turnovers[1] and merry-go-rounds,[2] things familiar in the plant today. Such pieces of equipment have been implemented in great numbers since the 1950s.

The many types of automation currently in the plant—other than the new type of robots or pieces we refer to as robots—really began in the 1950s. Walter Reuther was asked what the effect on the UAW membership would be as time went on. He said to keep in mind one thing, which holds true today as well: "Robots cannot buy cars and neither can people that are not working." There is a built-in security in our society. If people are not working, they do not have money and cannot purchase cars; therefore the companies can not make a profit. For this reason companies must face the fact that they have a social commitment to provide jobs, if for no other reason than to maintain their own profits. Reuther went on to negotiate contract provisions that addressed exactly those issues he predicted. The negotiated gains created jobs by allowing members time off while at the same time the plants could continue to be operated full-time. There would not be a drop in production but an actual increase, even though workers were spending less time in the plant.

CONTINUED PROGRESS

Succeeding UAW presidents, Leonard Woodcock and Douglas Fraser, continued in that direction, coming up with new types of benefits and proposals

1. A turnover is automation that takes a panel from a particular press or welding operation and turns the panel over for work to be done on the other side as it goes down an assembly line.

2. A merry-go-round is an automated section at the end of a line where large pieces, such as roof bodies or frames, are stacked on a circular, traveling conveyor and then are unloaded, as they turn around, onto racks or other types of storage devices.

to reduce work time and create jobs while at the same time providing security and increased productivity. In the 1970s automation took a leap forward, and new types of automation began to be used. Such equipment was mainly used for tedious, competitive, or dangerous jobs. The then healthy economy's demand for more and more cars hid or masked any effect of this increase in new technology in the replacement or displacement of workers, because there was a constant increase in the demand for cars. There was also a constant increase of new facilities. People were put in those new facilities to operate the new equipment, and generally automation was accepted just as it had been in the 1950s, as a benefit not only to management but to the worker as well.

CURRENT EFFECTS

In the 1980s the United States faces another stage in the evolution of automated equipment, one that is going to affect every area of the employment market requiring more advanced equipment. The Japanese invasion has taught us that we will have to automate if we are to remain competitive on a world basis. As a society we will have to address the fact that in their design, production, and inspection areas management will have to turn out a better quality product by designing a better quality product, and by allowing people on the floor to have more input on problems in production.

Union and nonunion members alike are now threatened by the advance of new technology. The recent decrease in car demand and the present shape of the economy have brought into focus the fact that as automation comes into the plant, people are going to be displaced. People may not necessarily be laid off as a direct result of automation, for there is always attrition in the work force—people retire or leave for one reason or another. But some people will be displaced, moved from their jobs. Displacement can move people into higher-paid jobs, jobs in semiskilled areas, but the negative side of the coin is that displacement can force people to seek jobs in other industries. Jobs are going to be created because of automation, but at the same time workers will face displacement within the plant and possibly in new industries, which may mean moving to other plants or newly opened facilities.

SUGGESTED SOLUTIONS

What are unions doing about the influx of automation? What are unions going to do about automation that can be reprogrammed without being torn out, simply by pushing buttons or changing a particular tape? The UAW is addressing these problems, and the following is a review of the recent resolution—involving much discussion of new technology—passed at the February 1982 Skilled Trades Conference, a policy-setting body within the union.

1. The UAW must insist that contracts assure workers' income and employment security.

2. There must be advance discussion before the implementation of any type of automated equipment.

3. The new technology designs must be subject to bargaining in order to develop systems that help workers rather than degrading them or placing them in secondary status.

4. Layoffs caused by the introduction of technological advance or change must be prohibited.

5. UAW craftworkers should be assigned to perform all the necessary skilled work.

6. There should be regular and periodic information meetings, informing members and keeping local representatives updated on developments in new technology. Much work has been lost to salary units in the past because either local leadership within the plant or international representatives at one time or another just agreed and allowed such work to go out, feeling that members were no longer qualified to do it. This resolution addressed that issue directly.

7. The use of computers must not be allowed for time study or disciplinary methods within the plant.

These are just some of the areas addressed by the Skilled Trades Conference. The Production Conference, another policy-setting body that met in February 1982, addressed the same types of issues, which are becoming critical and which in the years ahead the UAW will have to address.

Each succeeding contract since the 1930s has addressed the problems of the future. Subcontracting, productivity, job creation, and so on are not only benefits but also issues that affect UAW membership. Job loss and job creation benefits are not new; they have been with us all along. The resolutions outline some specific areas that must be addressed, but in addition the Skilled Trades Department of the UAW has staff members that attend various tool shows, seminars and colleges, and investigate developments in new technology. The UAW wants to know what new types of equipment our membership is going to be facing, and at the same time to be able to advise local unions on who should perform the resulting new types of work. The union should be well versed as to the type of tools that are going to be necessary in order to troubleshoot and maintain the equipment. When going into negotiations, the union can then assure that such equipment will be provided in the tool cribs, or that equipment will be purchased by the company for use by members or as replacements for tools being used now. There are also interstaff study committees throughout the international union that have prepared for upcoming negotiations and that address problems specifically in each area.

PREPARATION FOR THE FUTURE

The UAW is preparing to face the future. Automation is going to continue, but workers have to be prepared to share in the benefits it can bring. The process should not be one-sided; jobs will be lost to automated processes unless all concerned parties continue to address this issue. There will be fewer production facilities in the future. Moreover robots and optical sensing equipment are beginning to move into the next labor grade of jobs. In the past robots could perform only simple pick-and-place types of operations. Now optical sensing can actually inspect parts coming down the line, signaling a robot to reject certain parts and to stack, crate, or package other parts.

The totally automated factory, though controversial and an attention-getter, is not a reality. There will always be jobs that have to be performed, such as packaging, loading, and in some cases transporting; such jobs may eventually be of a much higher nature technically than those currently performed, but there will always be someone who has to troubleshoot the equipment. Will the trouble-shooters have to be all-around general mechanics, trained in both mechanics and electronics? Not necessarily; there will have to be teams of specialists on such jobs. We still believe that the individual craftsman will be needed in the future, and that is the direction we will be following as a union.

One problem with the coming of automation is the issue of security. Security can really be the only basis on which workers can continue to accept automation in the plant. The issue has been addressed in the past but will need special attention in the future. There have been temporary setbacks in current programs of paid personal holidays, which allow random time off and create jobs in the plant; but this is still a long-term union goal. The holiday program accomplishes several objectives: it allows management to continue to use the plant; it provides additional jobs because of those needed for replacement; and at the same time the union gets the benefit of added leisure time for its membership. Another issue for the future is lifetime employment guarantees: companies point at the lower wages of the Japanese, but they do not consider the fact that the Japanese are in many cases guaranteed a future with the companies they work for. Retraining for more highly skilled jobs is another necessity, along with protection of work that was at one time in unit.

THE SKILLED WORKER

A short article from a newspaper in Southfield, Michigan,[3] near the heart of the auto industry, outlines some of the problems facing the skilled work force, which must be addressed:

> People who create the miracles of modern industrial life are given little esteem by the society they serve. . . . How can it be that the kind of work that creates the hallmark of advanced society is scorned by the same society. . . ? The industrial revolution has changed our world. With the advent of the factory, a relentless process of deskilling began. Skilled craftsmen were a problem for factory owners because they could force higher wages simply by withholding their specialized essential knowledge from the workplace. Employers could end this dependence by breaking jobs down into tasks small enough that anyone could learn them, thus eliminating the need for craftsmen. . . . Without the need for special skills, any worker could perform any part of the assembly process with only a short training period. . . . Blue Collar workers . . . are not subject to such a measurement of output.

THE SEMISKILLED AND NONSKILLED WORKER

Not only will our society have to address the replacement or infiltration by automation of semiskilled or production-type operations; there is also going to be a management tendency to break down further some skilled classifications and skilled job operations, and wherever possible to assign these jobs to less skilled groups. There will have to be advance agreement by UAW members on installation. Design of equipment must take into account the safety element not only for those who work on the equipment, but also for those who operate it. Automated equipment must be safe in itself and easily shut off by an operator in case of malfunction. The equipment must be designed to conform to the human operating the equipment,

3. *Southfield Eccentric*, 8 Oct. 1981. Copyright © 1981, Observer & Eccentric Newspapers. Reprinted with permission.

rather than forcing the human to conform to the equipment—for example, to its speed of operation.

SUMMARY

Security of workers must be a top priority, safety must be assured, and there must be advance discussion and communication before the implementation of automation. Introduction of any automation must be timed so as to benefit workers. Natural attrition, rather than massive spur-of-the-moment layoffs, should be the only way of reducing the work force. Such a policy would share out the benefits a booming car market provides the companies: companies would still have to provide employment and job guarantees for people when the economy restricts the demand for cars.

Workers understand that the industry needs automation to remain competitive, but this must not happen at the expense of their livelihood. Workers must share in the benefits of new technology. Society as a whole must face the fact that for America to continue to prosper, people have to work. That consideration in and of itself is going to force America to face the responsibility of providing jobs of some nature. Those jobs may ultimately be work performed at home on a small computer unit tied to the workplace, or some other method, but there will have to be some way of providing employment for people if America is to continue to grow. That is a simple economic fact. The UAW is in the forefront, ahead of any other union in this country, in addressing these issues. As an international union the UAW plans to stay abreast of technological advances. The union has worked very closely with Carnegie-Mellon University and other universities in the Midwest on studies of automated equipment and its effect in the future. We have also maintained a close connection with the Society of Manufacturing Engineers/Robotics Institute of America. The UAW plans to continue addressing these issues, making sure that its members reap some of the benefits that the new types of automation will bring.

ANNALS, *AAPSS*, **470**, November 1983

Electronics and the Future of Work: The Impact on Pink and White Collar Workers

By DORIS B. McLAUGHLIN

ABSTRACT: Much as in the late 1950s and early 1960s we read about the impact that automation would have on blue collar jobs, so is literature now beginning to appear on what effects can be expected as new, highly sophisticated equipment is introduced into the office—the world of pink and white collar workers. Because of an expanding economy and a rapidly growing segment of the service sector, as well as the fact that the automated equipment introduced in the 1950s and 1960s was both unifunctional and expensive, the dire predictions as to its impact proved wrong. But the 1950s scenario no longer holds. Thus the introduction of new office equipment may well have an adverse impact on both the quantity and quality of jobs, particularly the lower paying jobs traditionally held by women. Hard data on the actual impact of the new technology on the office work force are hard to pinpoint: the phenomenon is a very recent one; other variables usually come into play; and we are now living in a global economy—jobs that appear lost may simply have been moved. Perhaps the best way to fill the data gap would be to work with one or more of the labor unions that have organized office workers to develop the best methods to study the subject.

Doris B. McLaughlin—Ph.D., University of Michigan, 1973—is an associate research scientist with the Institute of Labor and Industrial Relations at the University of Michigan. Her major work has been in the area of labor history and collective bargaining, with a recent emphasis on the impact of technology on jobs.

NOTE: The author would like to thank Audrey Freedman, Judy Gregory, Art Schwartz, Ted St. Antoine, and Harley Shaiken, as well as editor Robert Miller, for their helpful comments on an earlier draft of this paper.

THE major focus of this article is the technological revolution that both the so-called pink collar workers—clerical and administrative support staff—and the traditional white collar professional and technical workers are now facing.

It might be well first to look at what is known in general about the impact of technological change on jobs.

PREDICTIONS

The literature on the topic is abundant. Some articles, mostly reportorial in nature, praise the new technology. Their authors believe that it will take over the dull and routine, and free human beings to do more interesting and meaningful work. Quality of output will be higher, and labor and capital productivity will increase.

Much of the rest of the literature, however, is less optimistic in tone. These writers view the introduction of new technology as inevitable, but they point out the disadvantages, as well as the advantages, that it will bring. They see the possibility that more jobs will be lost than created and that the remaining jobs will be either more boring and routinized than they were before, or will require such a high level of skill that few will be educable to master them. Moreover any improvements in quality and productivity will in many cases come at the expense of human physical and emotional well-being.

Colin Norman[1] synthesizes much that has been written about the impact of electronic technology on jobs worldwide. Norman points out both the costs and the benefits involved in embracing technological change, but on balance he finds that not accepting it is the more disastrous route to take. Similarly Stephen G. Peitchinis of the University of Calgary has written a series of monographs on the problem of technology and jobs in Canada. Peitchinis's overall conclusion from the studies to date is typical of the cautious approach taken by the authors of much of the extant literature:

> Technological changes have both positive and negative effects on employment and individual skills. . . . Accurate anticipation of effects have proven difficult even in times of routine technological changes; it is doubly difficult in times of revolutionary changes.[2]

Harley Shaiken, now at the Massachusetts Institute of Technology, brings an added dimension to bear on the problem. As a one-time machinist and United Auto Workers' (UAW) Union member he has experienced the effects of technological change first hand. He has published numerous articles on the subject, and his latest book, tentatively titled *Work Crisis: Workers and Automation in the Computer Age,* is scheduled for publication some time this year. In his "Computer Technology and the Relations of Power in the Workplace,"[3] Shaiken views the impact of the new technology on jobs as three-pronged: its effect on (1) numbers of

1. *Microelectronics at Work: Productivity and Jobs in the Economy* (Washington, DC: Worldwatch Institute, 1980).

2. *The Introduction of Computer-Aided Design (CAD), Computer-Aided Manufacturing (CAM) Systems and Their Employment Implications* (Ottawa: Technology Branch, Industry, Trade and Commerce, 1980), p. 11.

3. IIVG Papers, 80-217, International Institute for Comparative Social Research, Berlin, Oct. 1980.

jobs, (2) the pace of work, and (3) their internationalization.

The first prong involves the whole question of the number of jobs that will be lost and the number that will be substantially changed. Like Peitchinis, Shaiken foresees both negative and positive effects, but he is generally pessimistic about what the impact on the number and kinds of jobs available will be.

Second, Shaiken sees the pace of work in the future largely controlled by machines rather than by the people who operate them, as more and more the two come to work side by side. The new technology also affords management the opportunity to monitor and control the work pace continuously, which can lead to a form of spying or stressful speed-up. Moreover, because the worker no longer sets the work pace, any discretion he or she had exercised as to how and when work is to be done in any given time period will disappear. The possibility of "schmoozing," a Yiddish term used to describe informal socializing, will be eliminated except during formal breaks, and social and physical isolation will be added to the disadvantages of a job that already can be mind numbing. Ironically this could redound to the employer's detriment: much sharing of information on how a job can be done more effectively takes place in the course of informal conversations between fellow workers.

The third factor Shaiken notes is the possibility of the internationalization of production. He sees this as a particularly ominous threat to organized labor, since in the event of a strike threat in one area, management can simply move production someplace else:

> If an unexpected labor dispute holds up tooling scheduled to be built say in Germany, the job can be completed in England or Spain by transmitting all the relevant data to numerically controlled machines in those countries.[4]

To Shaiken the key issue is, Who will control the new technology and for what purpose?

While he notes the threat to labor, Shaiken also points out that such "vast concentrations of computer power in relatively few hands" might prove to be an Achilles heel to management.[5] That prediction was accurate. In May of 1981, eight months after Shaiken's article appeared, the British data processors and operators who run the big computers that process the country's value-added tax went on strike. According to a report in the *Wall Street Journal* that strike made "a shambles of the country's revenue collections, interfering with defense operations, and, among other things, preventing payments to dentists in Scotland."[6] In a subsequent article[7] Shaiken did note that as long as management has the authority to redesign work patterns the impact of that Achilles heel can be countered. It was the highly centralized nature of the computer system that accounted for the strike's success. A more decentralized system would have permitted management to transfer work readily from one site to another, at least mitigating if not eliminating the strike's adverse effects.

4. Ibid., p. 55.

5. Ibid., p. 58.

6. Barry Newman, "Computer Strike Snags a Bureaucracy," *Wall Street Journal,* 19 May 1981, p. 27.

7. "Computers as Strikebreakers," *Technology Review,* Apr. 1982, pp. 50-51.

PAST FORECASTS VERSUS THE PRESENT SCENARIO

It should be noted that in the late 1950s and early 1960s we were reading much the same thing about the new technology—then in terms of the impact that automation would have on employment. Then as now a few writers predicted that the new technology would be a blessing. It would do work that people did not like to do—the dirty, the dangerous, the tedious. Moreover it would create more jobs than it would eliminate. Most of the literature, however—then as now—was less sanguine in tone. For many workers, it was said, automation would be a curse, not a blessing. It would result in massive displacement, and the redundant workers would have no place to go.

Fortunately the optimists of those days were correct. Those dire predictions did not come to pass. An expanding economy and a rapidly growing segment of the service sector—transportation and public utilities; wholesale and retail trade; finance, insurance, and real estate; such miscellaneous services as health, hotel, auto repair, and legal service; and all levels of government—absorbed both those displaced by automated equipment and the new entries into the labor market. Moreover those early machines were unifunctional—and thus subject to rapid obsolescence—and expensive. As a consequence they were not introduced as rapidly or extensively as they had envisioned.

Today the 1950s scenario no longer holds. While some industries are economically healthy and can absorb workers displaced by the introduction of technological change, those that traditionally have fueled the economy are not. In addition, because of the advent of the microprocessor in the early 1970s, the equipment now available is multifunctional, easily reprogrammable, and reasonably priced compared to the past. Moreover, while workers in the service sector now represent a larger proportion of the labor force than those holding jobs in goods-producing industries, with the exception of the expansion of available work in the trade and health-care fields, the numbers of jobs in the rest of the service sector, both private and public, are either stagnant or shrinking.[8]

Thus, because of the current economic climate coupled with the advent of more sophisticated technology and the changes taking place in the service sector, the dire predictions about the impact of technology on jobs may now prove to be accurate. They may not prove true in the long run—by 1994 we might be feeling the effects of the baby bust—but the short-run ramifications should concern us all.

It is well to remember that not all of the service sector was expanding in the 1950s and 1960s. Some segments were already automated or in the process of becoming so. For example, consider the retail food industry. The advent of the automobile and home refrigeration had helped make the self-service supermarket a reality, and had changed the composition of the labor force in that industry dramatically. The ice man was gone. The ranks of the store clerks kept dwindling. Soon the milkman all but disap-

8. See, for example, Alfred L. Malabre, Jr., "Service Jobs Keep Expanding in Recessions, Make Up Ever-Larger Share of Work Force," *Wall Street Journal,* 15 Jan. 1982, p. 40. The title of the article does not give a complete picture of its contents. Malabre also points out how the introduction of automated equipment is affecting the service sector as a whole.

peared. The technological revolution in that industry is still continuing, of course. The Universal Product Code and attendant computerized scanner, as well as boxed beef and other prepackaged produce, are all leaving their mark on the size and composition of the supermarket labor force. As an aside, note that in the name of convenience almost every technological innovation mentioned has transferred work formerly done by a paid employee to a new team: the supermarket customer and a machine. The same phenomenon can also be seen in other service sector industries that have become automated. Phone installation, direct dialing, self-service elevators and check cashing are just a few additional examples.

The segment of the service sector that was expanding in the 1950s and 1960s involved office work. Both so-called pink collar jobs—lower-paid clerical and administrative support positions often held by women—and the traditional higher-paid white collar professional and technical jobs—usually held by men—were on the increase. The growth was particularly evident in the public sector, at all levels of government. The expanded opportunities to work in government and other offices absorbed those displaced by automation, as well as the new entries into the labor market.

That route, however, is no longer open. Government employment at most important levels is shrinking. Equally important, the office is undergoing a technological revolution of its own. The introduction of the computer, the word processor, and other so-called smart machines, guided by microprocessors, is already having an impact on the number of people required to do the work, the nature of the work performed, and the time spent doing it. Thus the pink and white collar segments of the labor force are now face-to-face with the new technology, and are beginning to realize its potential for both good and ill.

Many forecasters are now predicting that the new, sophisticated technology will most adversely affect the pink collar clerical jobs, mostly held by women.[9] They envision a computer terminal on every managerial desk, permitting electronic direct entry and subsequent transmittal, storage, and retrieval of all intra- and interoffice reports, letters, and other messages. Ultimately there will be little need for typists, file clerks, or secretaries on the office staff. Routine or repetitive word processing and data entry that the managerial, sales, and other professional office staff still on board find too tedious can be given to someone—again usually female—working alone in a computer-equipped home, paid on a piece rate basis. Alternatively the work can be sent via satellite to be processed overseas, in some low-wage country like Barbados. Both phenomena—the new version of a cottage industry at home and what Harley Shaiken calls "telescabbing" abroad—are already in evidence.[10]

Even the more optimistic among the prognosticators believe that the office of the future will be far less labor intensive than it is at present. The cutback they

9. See, for example, testimony given on 23 June 1982 by Judith Gregory, Research Director, 9 to 5, to the Subcommittee on Education and Labor, U.S. House of Representatives, p. 3.

10. Some experts predict that by the mid-1990s as many as 15 million workers will earn their primary income from homework. See "If Home is where the Worker Is," *Business Week*, 3 May 1982, p. 66. Shaiken is quoted in Andrew Pollack, "Latest Technology May Spawn the Electronic Sweatshop," *New York Times*, 3 Oct. 1982, sec. 4, p. 17.

envision can be as high as 50 percent. As an offsetting factor they assert that the employees who do remain will have a higher quality of work life.[11]

Based on what we already know about the impact of technological change on jobs, that assertion is too optimistic. Presently—and there is no reason to believe that that will change—the impact of the introduction of new technology varies. Some of the remaining employees do in fact have more meaningful and interesting work. The rest, however, become little more than bored button pushers.

But will jobs be created as well as lost? It depends on what segment of the service sector is being studied. Of course the microelectronic equipment itself has created the need for new jobs, not just to build, install, maintain, and repair it, but to run and program it, and to check and analyze its output. The traditional service industries are also affected, however.

In banking, for example, jobs are created once microelectronically controlled equipment is installed. With it in place, banks can provide services to their customers, such as billing services, that were formerly unavailable because those services had been too expensive, time consuming, or cumbersome to do manually. As a consequence employment in banks in this country grew by 50 percent between 1970 and 1980.[12]

The other side of the coin, however, is that the 50 percent increase in available bank jobs may not have benefited the human teller displaced by an automatic machine, or the typist no longer needed because a word processor has been installed. Not all banks have retraining programs for the higher skilled jobs that are created. Many bankers prefer to hire someone new, already trained to operate the new equipment. Even if his or her employer does have an in-house retraining program, the redundant worker may not have the educational background—or the incentive—to take advantage of it.

In other segments of the service sector some new jobs have been created for middle management, but not enough to offset the number of nonmanagement jobs lost. The telephone industry serves as a good example of that phenomenon. When computerized equipment that permitted direct dialing, automatic systems to record calls, and machine-controlled maintenance and repair of equipment were installed, data analysts were added to the staff. But the number of new nonmanagement hires was much smaller than the number of telephone operators and skilled linemen who were displaced.[13]

A dramatic case in which new equipment has apparently resulted in a net loss of jobs is the building and construction industry, specifically the architect's

11. See, for example, Vincent E. Giuliano, "The Mechanization of Office Work," *Scientific American,* Sep. 1982, p. 163.

12. Martin L. Ernst, "The Mechanization of Commerce," ibid., p. 138.

13. According to Communications Workers of America President Glenn E. Watts, in one year alone—1977—operators' jobs declined by 10,500 positions. See his "Technology, Telephones and Jobs," in *Silicon, Satellites and Robots: The Impact of Technological Change on the Workplace,* ed. Dennis Chanot and Joan M. Baggett (Washington, DC: Department of Professional Employees, AFL-CIO, 1979), p. 30. According to Robert Howard, between 1974 and 1980 craft employment at American Telephone and Telegraph dropped by almost 22,000 jobs. See his "Brave New Workplace," *Working Papers,* Nov.-Dec. 1980, p. 22.

office. One architect using a computerized library of preengineered components can produce in one afternoon building plans that formerly took three months to complete. There is no longer a need for a draftsman or an engineer. Computer aided design (CAD) also reduces the number of architects a firm requires. For example, one firm that had once had 12 architects on its staff now requires only two, although that firm's revenue has quadrupled. Nor is there a need for someone to estimate the amount of building materials needed, to make out purchase orders, or to determine what material should be available at the building site. The computer had been programmed to assume those duties. Mistakes are corrected before the building is ever begun, and downtime on the site itself—a major factor in driving costs up—is largely eliminated.[14] The new process results in enormous savings in time and money but bodes ill for the newly graduated architect, draftsman, or civil engineer looking for a job.

BUFFERING THE IMPACT

What appears to be happening in the pink and white collar worlds is the development of a two-tiered society. In fact according to Robert Howard the two-tiered society of the future is already a reality at American Telephone and Telegraph. In contrast to many of the other employers of pink and white collar workers, that company is unionized; and it is the union members who are bearing the brunt of the new technology's impact:

> on one side of the divide, highly skilled work controlled by non-union management personnel; on the other, degraded computerized work relegated to the unionized workers of the CWA [Communications Workers of America] bargaining unit.[15]

While a union can bargain for contractual language aimed at affording some protection to the adversely affected members, that phenomenon—leaving only unskilled, easily replaced, workers in the unionized segment of an organization's work force—nevertheless substantially erodes the union's power base as an institution. The wider implications for the American labor movement as a whole should be clear.

Thus those with the necessary education and training—the upper tier—may well benefit from the new technology, whether it be the librarian or lawyer using an information retrieval system or the doctor consulting Internist I, a programmed diagnostic tool. Their quality of working life may be lessened in the sense that their support staff has shrunk and they are accessing the new equipment directly. But in the sense that they have so much more information readily available to them, their quality of professional work life has been enriched.

What of the lower tier, those with routinized, dull jobs? Studies have shown that once work is routinized, turnover increases markedly. To combat turnover some employers have instituted work-sharing or job rotation and job enrichment programs, or have switched from the traditional to flextime work patterns. Others are providing clearly stated career paths. None of these changes has been a panacea, but where they have been successful the employer has benefited in terms of both higher productivity and better employee

14. "CAD Redraws the Architect's Job," *Business Week,* 15 Mar. 1982, pp. 134, 136.

15. Howard, "Brave New Workplace," p. 25.

morale. Both part-time and flextime work patterns may become even more common in the future, simply because those work modes suit the needs of the young, involved in work-study programs; the old, no longer interested in full-time employment; and the female, with home and family responsibilities.

Four years ago California, in connection with its unemployment compensation plan, instituted an interesting work-sharing scheme. Under that program, the employee's time and salary are cut by 20 percent, and the state pays him or her—and mostly it has been her—one-fifth of the unemployment benefits he or she would have received if totally unemployed. The program has advantages for all concerned: the employee still has a job and continues to be entitled to fringe benefits; the employer retains a trained work force; and the state saves some money. Because the California plan has been so successful, Arizona and Oregon have each since passed the same legislation, and other states are considering it.

Labor unions, whatever the color of the collar of the workers they represent, have been in the forefront in directing their attention to the human costs involved in adjusting to new technology. While unions generally have been willing to accept the attendant changes, they have at the same time attempted to negotiate contract language that helps to make those human costs less painful. It might therefore be useful to look at what they have been bargaining for.

Among the clauses that unions—the American Federation of State, County and Municipal Workers (AFSCME), the Service Employees International Union (SEIU) and the Communications Workers of America (CWA) among them—have sought to have included in their labor agreements are:

—a requirement that the union receive advance notice when the employer intends to introduce new technology;

—rate-retention obligations, should the newly-structured job require less skill than it formerly did;

—early retirement provisions;

—employer-financed training and retraining rights;

—interplant and intraplant transfer rights, often coupled with relocation pay where applicable;

—the establishment of a joint labor-management committee, to deal with, among other things, health and safety problems;

—a severance pay provision;

—a restriction on time-motion monitoring or production quota setting;

—spread-the-work provisions, such as longer vacations, more paid personal leave, shorter hours; and

—a provision that reduction in force due to the introduction of new technology will be by attrition or reassignment only, for those in place when the new technology is introduced.

With one exception, none of these provisions can be said to offer complete protection from the adverse impact of the introduction of new technology, but they all do constitute at least partial cushions. The one exception—requiring reduction in force by attrition or reassignment only—does offer job security to those covered by the contract when the new technology is introduced, but it does nothing to help the would-be new entry. Moreover as far as the union itself is concerned, a dwindling membership is virtually guaranteed.

RETRAINING

If the two-tier effect described is to be avoided, we should also take a close look at the whole broad field of education and training. To use the example of the unemployed teller or typist who would like to be transformed into a computer programmer, that process will ordinarily take about six to nine months, if the trainee is to have enough training to be employable at the entry level. The paramount question then becomes, Whose responsibility is it to support the trainee during the period—the former employer's, the prospective employer's, if different, the government's, or the trainee's?

Ideally the trainee should have had sufficient background, as a part of his or her general education, to fit into the new position with a minimum of training, but that is rarely the case. Our educational system does not usually prepare students to deal effectively with a cluster of skills. In fact employers often complain that the system does a mediocre job of preparing the student for the world of work at all. The complaint is the same whether directed at the ill-prepared high school graduate, college graduate, or freshly-minted master of business administration. Indeed a common fear expressed by the prognosticators is that because of the inadequacies of our educational system, the new technology that emerges in the future is less important than the fact that we will lack people who know what to do with it. In other words they predict a shortage of upper-tier people, and a glut of those in the lower tier.[16]

We could learn from one organiation that does make an effort to prepare its graduates to function in the emerging electronic world, the Organization for Rehabilitation through Training (ORT), a Jewish agency that functions worldwide. Originally founded in the early twentieth century by wealthy Russian Jews seeking to provide vocational skills to their less fortunate coreligionists, ORT has evolved into a far-reaching training organization that, at least in its model schools, teaches high school youth a variety of skills. For example those students seeking to enter the computer field learn not only how to build, maintain, and run a computer, but also how to program it. As a consequence graduates of that and ORT's other programs readily find work.

A past president of World ORT, octogenarian William Haber, many years ago voiced another principle the organization espouses, and one that seems even more relevant today: the concept that education is a continual process; that after a time one's initial training is outdated; and that one must take care to keep abreast of changes in the field on a continuing basis.[17]

16. See, for example, "The Speedup in Automation," *Business Week,* 3 Aug. 1981, p. 67.

17. "Authority Without Freedom: The Birth of the Teach-In," *Michigan Quarterly Review,* Oct. 1968, pp. 262-67.

CONFRONTING THE RESEARCH GAP

Much of what we know about the impact of the introduction of the new, highly sophisticated, technology—the effect of the advent of the microprocessor—on pink and white collar jobs is sketchy and imprecise. One reason for the lack of specificity is that it is such a recent development. The microprocessor came on the market only in 1971, and then it was used primarily by the military. Only in 1973-74 did the focus begin to shift to civilian use. The widespread introduction of office equipment fueled by microprocessors is of even more recent vintage.

Moreover hard data on the number of jobs lost, created, or changed solely—or even primarily—because of the introduction of new technology are hard to pinpoint. Many other variables usually come into play: changes in consumer preferences; intraindustry or interindustry competition worldwide; altered economic conditions; industry restructuring; new legislation; government regulation or deregulation; and so on.

Beyond that, given the global economy we now live in, jobs that appear to be lost in one location may not really have disappeared at all. They may just have been moved.

Much of the writing on job impact that does deal with numbers describes what firms that are installing the new technology, or those that manufacture it, estimate will happen by 1990 or the year 2000. In-depth studies of what has already happened, weighing all possible variables that have had an impact on specific jobs and individual workers, are lacking.

Perhaps the best way to try to fill that gap would be to work with one or more unions that have organized pink or white collar workers—the organizations most familiar with what is actually happening on a day-to-day basis—to develop the best methods to study the subject. With their help a series of in-depth studies could be designed, each limited to a cross-regional sample of, say, office workers in one specific industry that has installed new technology.

The questions to be addressed should include the following:

1. In general. What factor or factors influenced the decision to introduce the new technology—for instance, outside competition, and so on?

2. Concerning the job. How have jobs changed? How many have been lost or created? How much work, if any, that had traditionally been done in the office has since been transferred elsewhere; where were they moved to; who is doing them; and has the wage rate changed—up, down, the same? Did factors other than the introduction of the new technology come into play with regard to the change in jobs? If so, what were they, and how much impact did they have, as opposed to the new technology?

3. Concerning the worker. What happened to those who were displaced? Were they trained, and by whom, for other jobs in the industry? In the same or different location? Who bore the cost of a transfer if one was necessary? If they were not retrained, what happened to them subsequently? Who was hired to do whatever jobs were created? What

has been the impact on the affected workers, in terms of skill level, job satisfaction, work environment, relationship to family and community, and the like?

4. Concerning the union. What has the union tried to do—either informally, through collective bargaining, or by legislative lobbying—to alleviate whatever adverse impact the introduction of new technology has had on its members? How successful has each of those efforts been, and why? What have unions learned about labor organization action in other countries that could usefully be implemented here?

A series of such in-depth studies and the resulting analyses would give us a clearer picture than we now have of the problems encountered, and the solutions available to confront them adequately, when new technology enters the pink and white collar world.

ANNALS, *AAPSS*, **470**, November 1983

Technology and Employment in Telecommunications

By RONNIE J. STRAW and LOREL E. FOGED

ABSTRACT: The nearly 700,000 members of the Communications Workers of America have witnessed first-hand the effects of technological change on jobs and workers. Although technological change can bring many benefits to Americans at home and at work, if implemented improperly technology can eliminate entire classifications of jobs while deskilling and dehumanizing the jobs that remain. Worker participation in all stages of the process of technological change can provide added perspective and hands-on expertise, benefiting both labor and management by contributing to the design of challenging and productive jobs that enhance economic efficiency and growth.

Ronnie J. Straw has been director, development and research, Communications Workers of America, since 1970. Prior to that he was assistant executive manager and staff economist, National Telephone Cooperative Association. He did Ph.D. work in economics at George Washington University, completed an M.A. in economics and public administration at Syracuse University, and earned a B.A. in economics and business administration from Geneva College.

Lorel E. Foged is an economist/statistician, development and research, Communications Workers of America. She earned an M.A. in economics from George Washington University, 1983, and a B.A. in economics from Franklin and Marshall College, 1978. She coauthored with Mr. Straw "Job Evaluation: One Union's Experience," Cornell Industrial Labor Relations Report *(Spring 1982).*

TECHNOLOGY seems to have become the theme of the 1980s. We hear daily of the wonders of technological change in store for us both at home and at work in the years ahead. The media have glamorized and fantasized about the imminent joys and tragedies of the microelectronic age, enthralling—and sometimes threatening—Americans with the inevitable transition of lifestyles that, the articles insist, will leave no man or woman unaffected.

Technological change is not unfamiliar to the nearly 700,000 workers represented by the Communications Workers of America (CWA) who have experienced first-hand the rapid introduction of new technology into the workplace. The systematic permeation of new technology into every aspect of work has made technology a key issue for our union. The history of the telecommunications industry in the United States has been characterized by technological change, enabling the provision of telephone service to over 90 percent of all households. Our members are the first to recognize that technological advancements can provide a host of benefits to our nation.

If implemented correctly, technology can lead to decreased costs, new jobs, and improved productivity, supplying a much needed boost to our lagging economy. New technology can help raise our standard of living, enabling us to enjoy a greater number of quality products and services. Technology can bring us everything from time-saving gadgets for the home to life-saving medical devices and can even prevent the world from running out of energy and natural resources. In the workplace, developments in technology can minimize or eliminate the dull, monotonous, and unsafe aspects of work and can open up new employment opportunities.

Our eagerness to reap the benefits of technological change must never blind us to its possible costs, however. Just as the industrial revolution moved workers from the farm to the factory, and the post-World War II boom transferred work to the service sector, the impending implementation and dispersion of the microprocessor threatens to fragment, deskill, and dehumanize work in all sectors. All too often businesses plan and promote new technology in a shortsighted attempt to decrease costs, thereby raising profits and shareholder incomes, paying little or no attention to the effects of innovations on workers.

Technological changes at the workplace can reduce skill requirements, fragment job content, foster both mental and physical stress, create new safety and health problems, and eliminate entire classes of jobs. Machines once merely aided workers in their jobs, but high-capability microprocessing equipment now monitors and controls workers, dehumanizing the workplace. Employees feel like mere appendages to the machines, not like individuals.

THE UPS AND DOWNS OF TECHNOLOGY

Without careful foresight and planning, the adverse effects of technological change can be so overwhelming as to overshadow or even negate the desirable benefits of innovation. The Bell System, which employs about 600,000 of our members and has been touted as part of the leading high-tech, rapidly growing telecommunications sector, has had its share of technological disappointments.

Prior to negotiating a national agreement with the Bell System in 1980,

CWA engaged a noted economist and productivity expert, Peter Chinloy of the University of British Columbia, to study labor's contribution to productivity in the Bell System. Labor productivity has been growing at an accelerating rate for years in the Bell System, with an average growth rate of 7.4 percent per year in the 1970s, compared to 1.3 percent for the total private U.S. economy. Total factor productivity grew by 4.9 percent per year in the 1970s, compared to only 1.0 percent for the total economy. The Bell System has consistently contended that this rapid growth has been due to technology. Our study, however, showed that economies of scale, not technology, have caused the high productivity growth, and labor has made a substantial contribution to this growth, contributing roughly 2 to 3 percent, or about 30 percent of the total output growth per year.

We found that as new technology has been introduced into the Bell System over the years, the effect has been sometimes to decrease and sometimes to increase costs. The net effect over the period we studied (1947-78) has been just about zero. Thus technological change in itself, defined as the shift in total cost over time, at the Bell System has had little substantive role in the industry's high productivity growth. Instead it is the hard work of labor together with other inputs that have enabled the Bell System to take advantage of increasing returns to scale and achieve high productivity.

Our results, though surprising to many, were not unique. Three recent studies confirm our conclusion that it was economies of scale, not technical change, that has caused rapid productivity growth in the Bell System. First, economists Michael Denny, Melvyn Fuss, and Leonard Waverman conducted a productivity study at Bell of Canada and came up with similar conclusions. They found that technical change was responsible for only about one-fifth of the productivity growth, whereas economies of scale accounted for about three-fifths. The remaining one-fifth resulted from the use of non-marginal cost pricing.[1]

Second, in a study of the U.S. Bell System conducted by M. Ishaq Nadiri and Mark A. Schankerman of the National Bureau of Economic Research, it was found that economies of scale accounted for 60 to 70 percent of total factor productivity, although this is declining. They found, in concurrence with our study, no statistically significant downward shift of the cost curve reflecting technical change. In other words technology has had a neutral effect on productivity.[2]

Third, a study of the U.S. electric power industry by Frank M. Gollop and Mark J. Roberts of the University of Wisconsin showed that technical change actually caused a reduction in marginal productivity of the fuel input so strong as to outweigh any positive, cost-reducing effects on productivity contributed by the labor and capital inputs. The net effect of technological change was that productivity in the electric power indus-

1. "The Measurement and Interpretation of Total Factor Productivity in Regulated Industries, with an Application to Canadian Telecommunications," in *Productivity Measurement in Regulated Industries,* ed. Thomas G. Cowing and Rodney E. Stevenson (New York: Academic Press, 1981), pp. 179-218.

2. "The Structure of Production, Technological Change, and the Rate of Growth of Total Factor Productivity in the U.S. Bell System," in ibid., pp. 219-47.

try was 2.4 percent less than it would have been in the absence of technical change.[3]

One plausible explanation for this unfortunate outcome of technology at the Bell System is that management introduced technology into the workplace without regard for the human factor. Because labor and capital work together in relatively fixed factor proportions at Bell, any rapid changes in technology would upset this balance and restrict, or even negate, productivity growth.

UNION INVOLVEMENT IN THE PROCESS OF TECHNOLOGICAL CHANGE

Unions have long been perceived as deterrents to progress, seeking to block technological change to attain their self-serving goals. Not only is this perception naive and inaccurate; it is illogical from the point of view of workers. Union members take pride in their work and welcome technology that enables them to broaden their job knowledge and enhance their technical skills. Unions are also the first to promote a profitable, financially sound company, for when the employer does well, we do well too. The problem arises when workers whose jobs have been downgraded or eliminated by technology observe corporations spending billions of dollars acquiring their competitors or sending work overseas, while low-paid or unemployed workers cannot afford the necessities of life for themselves or their families.

At CWA we believe that a nation as rich in knowledge and natural resources as ours can achieve the benefits of technological change without sacrificing human needs. The key to accomplish this, we believe, is to ensure effective union involvement in all aspects of the technological process, from inception through implementation. Labor is our nation's most valuable resource, and we think we should use it efficiently and properly.

Our union has directly addressed issues of technological change for over a decade. In the late 1960s we became increasingly aware that our members were struggling with various types of job pressures. In 1969 and 1973 we conducted two major surveys to determine the scope and nature of these job pressures and were surprised by the extent of the problems our members faced. Many were troubled by oversupervision and control, compounded by a general feeling that managers did not trust workers to perform their jobs responsibly and competently.

These problems were aggravated by the systemwide introduction of computerized equipment, such as the electronic switching system (ESS) in the 1970s. These systems have drastically transformed the nature of work in the Bell System, simplifying some jobs and complicating others. For example, some formerly skilled craft techniques have been reduced to plugging in and pulling out modular equipment, while certain clerical functions have been intensified by machine controlled work speedups and required knowledge of complex coding systems.

The most significant job pressure arising from technological change had been management's growing ability to monitor and control workers from centralized locations away from the work sites, robbing workers of autonomy, human contact and dignity.

A familiar example is the telephone operator, who no longer plugs calls in

3. "The Sources of Economic Growth in the U.S. Electric Power Industry," in ibid., pp. 107-43.

and out of the old cordboard. Instead operators plug themselves into a console that is part of the traffic service position system, (TSPS). This computerized system, which has eliminated local phone offices by centralizing work locations, channels calls to the operators, predicts future call flows, computes and catalogues each operator's average time spent per call, and even schedules work breaks. Because the computer schedules virtually no pause between incoming calls—which have doubled or in some locations even tripled in number to between 80 and 120 calls per operator per hour—operators are closely confined to the equipment, causing undue physical and mental fatigue. Considering too the fact that operators must now respond to customers with scripted phrases and sentences, technological change has devastated their work, removing most decision making and human response from the job.

The CWA has consistently made excellent gains for our members through collective bargaining, and in the 1970s we began to utilize this traditional process to tackle the issues of technology and job pressures. We succeeded in winning several substantial national improvements in this area, including

—upgrades for the operator and service representative classifications, both of which had been significantly affected by technological change;

—a letter of agreement to eliminate diagnostic remote monitoring;

—a letter of understanding protecting traditional telephone work from outside contracts;

—a reassignment pay protection plan to maintain wages for long-term employees who experience technological downgrades;

—a technological displacement clause to enable workers to opt for termination pay rather than be transferred to a lower-rated job;

—a supplemental income protection plan to maintain incomes for workers declared surplus because of technology;

—a joint quality of work life committee and a joint job evaluation committee; and

—a network of joint technology change committees, and provision to CWA of six months' notice of any major technological change.

Our 1983 bargaining resolution featured the key issue of employment security, a concept that recognizes the lifetime value and potential of the worker. In addition to expanding on the previous items, we also listed as goals the formation of a joint national technology change committee, a jointly managed career development and training program, and provision for union participation in the conception, design, testing, and implementation of new technology.

Despite these substantial gains we saw that job pressures were still all too prevalent in the workplace. The rapid introduction of technology created problems that needed to be dealt with on a day-to-day basis on the shop floor. The quality of work life (QWL) process has made significant headway in dealing with daily work site problems. Since the joint QWL committee began its work in late 1980, CWA and American Telephone and Telegraph have jointly agreed on a set of objectives for working together in cooperative problem solving. We found that although the traditional adversarial relations work well in the collective bargaining process, many of the local workplace difficulties can best be resolved jointly. Nearly 40,000 union and management persons have been jointly trained around the country. About 450 of those persons, roughly half union and half management, are working as group facilitators for the 800 active joint local work force groups. Training has occurred at all levels of CWA and American Telephone and Telegraph. The rapid success of our QWL process can be largely attributed to a sincere effort by those involved to develop mutual trust and respect to solve problems of joint concern.

Although CWA is pleased with the progress made in QWL efforts, QWL is limited in that it deals with the effects of technological change but does not address the causes. Joint technology change committees (TCCs) were set up in each Bell Operating Company, Western Electric, and Long Lines in a first attempt to involve the union in the process of technological change. These TCCs consist of not more than three company and three union representatives, usually including an assistant vice-president from the company and a bargaining committee chairman from CWA. During the 1980 contract the TCCs dealt mainly with the application of existing contract language. We expect these committees to mature and expand the scope of responsibility and influence in the years ahead.

STRATEGY FOR THE FUTURE

The assumption that technology will remove the drudgery from our lives and afford us more leisure time is alluring, but unfortunately there is no guarantee. We have already seen the effect of technology on jobs in the healthy telecommunications industry, and this example does not conform to the rosy outlook. In addition to affecting job content within the firm, technological change will have significant effects on total employment on a macroeconomic level. Three possible scenarios could arise over the long run.

1. Ideally technology will decrease costs, raise productivity, increase real wages, and create new, satisfying jobs. The only unemployment will be short-term frictional fluctuations with no long-run involuntary unemployment. This scenario presumes a full-employment, competitive market economy with an equitable income distribution.

2. A more pessimistic view would feature gradual increases in long-term structural unemployment as technology replaces and deskills jobs. These structural problems arise from market failures in the allocation, valuation, and mobility of resources. The imbalances could be corrected by deliberate governmental action and encouragement to the private sector, especially to promote training and retraining of workers in the economy.

3. The final and most formidable scenario foretells continuing long-term involuntary unemployment. The technology replaces more labor than can be utilized even after retraining and employment programs are conducted. Such a rise in the so-called natural rate of unemployment would lead to declining real incomes, stagnating or even falling productivity growth, serious economic imbalance, and a decline from our position of global leadership.

The first scenario is at best highly unlikely given our present economic situation. Although the final outcome will probably be closest to the second scenario, we may drift dangerously near the third if policymakers persist in their present economic, political, and social policies, which fail to take the human factor into account. In any case the choice is ours. Technology is a matter of social choice; it is not driven by some elusive force inherent in the process of technological change.

Some variation on the second scenario is most likely, although solutions to the actual technological unemployment will not be simple or quick moving. It has been pointed out that according to employment projections by the Bureau of Labor Statistics, the largest employment growth caused by new technology will be in low-skilled work. Although the fastest growth rates will be in high-tech areas such as computer operators, analysts, and mechanics, the absolute increase in these job categories will be small relative to the increase in low-skilled jobs. The five occupations expected to add the greatest number of new jobs by 1990 are janitor, nurse's aide, sales clerk, cashier, and waiter and waitress.[4] Even if targeted training programs effectively prepare and provide workers with employment opportunities, the problem of underemployment remains to be tackled.

Since technological change is a reality for CWA members, our union has made our own plans for dealing with innovations in the future. The key to success—which for us means a healthy, profitable company with employment security and satisfying, fairly paid jobs for the workers—is effective union involvement in all stages of the technological process.

In the short run we hope in the 1983 national negotiations to attain additional contractual rights regarding technological change, as already noted, especially by expanding the activity and responsibilities of the technology change committees, both formally and informally. We are also considering a training program for union members, and possibly managers too, to introduce them to the concepts of the technological change process.

Our mid range goals include continued extension of bargained rights in 1986 negotiations and greater participation in the initial conception, design, and implementation stages of technological change. A method that has worked in other industries is use of a slice team—a task force composed of experts from all levels of the business, from shop floor on up—to analyze and make recommendations on a particular piece of technology. The QWL work force teams could then be tied in with the TCCs to deal with any problems that remain once the technology is implemented.

We also expect continued support of national legislative initiatives to sanction and lend legitimacy to our efforts, as is done in some European nations. Areas of support could include a national reindustrialization policy, training and retraining programs, certain tax policies favoring joint technology endeavors or employment growth, and requiring social, economic, and employment impact statements from businesses contemplating major technological or organizational changes.

In the long run we expect union participation in all aspects of technological change to be an ongoing process. Union

4. Henry M. Levin and Russell W. Rumberger, "The Educational Implications of High Technology" (Unpublished discussion paper, Stanford University, Institute for Research on Educational Finance and Governance, Stanford, CA, Feb. 1983), pp. 4-6.

involvement will ensure that appropriate consideration is given to the human element. We also feel that workers, who are the only true experts on the jobs they perform, can supply valuable input into the developmental and testing aspects of the process, thus preventing many of the technical inefficiencies that restricted productivity in the past. In the past, decisions involving technology have been the sole domain of managers, many of whom are too far removed from the workplace to make sound judgments on new technology. In other cases, especially in large bureaucracies such as American Telephone and Telegraph, managers may be distracted or prevented from making optimal decisions because of regulatory restrictions or pressures to increase shareholders' or their own incomes. While the choices made by the managers may be cost-efficient in the short run, they may differ greatly from the social optimum.[5] Union involvement will give perspective to the process of technological change, benefiting both workers and management by contributing to the design of challenging, productive jobs that enhance economic efficiency and growth.

5. Recent economic models of bureaucratic behavior have focused on the tendency for managers to maximize more than just profits. See W. Niskanen, "Bureaucrats and Politicians," *Journal of Law and Economics,* Dec. 1975; and O. E. Williamson, *The Economics of Discretionary Behavior: Managerial Objectives in a Theory of the Firm* (Englewood Cliffs, NJ: Prentice-Hall, 1964).

ANNALS, *AAPSS*, **470**, November 1983

The Artificial Alien: Transformations of the Robot in Science Fiction

By MORTON KLASS

ABSTRACT: The robot, though a relatively recent arrival in the real world, has been a subject of interest in literature—primarily, though not exclusively, science fiction—for most of this century. It is possible to note significant transformations in perceptions of, and attitudes toward, the robot during this period by reviewing the literature. Thus the earliest robots are flesh and blood creations, and are perceived as potentially inimical in a number of ways. Over time the robot becomes a mechanical equivalent of humans and takes on many but not all of the attributes of aliens: interestingly the one most threatening alien attribute in Western perception—that of sexual threat—is not accorded the robot. Instead the robot is perceived as the perfect and perpetual servant, though with a distinct potential for danger. It might make humans superfluous in certain areas, but if humans can manage to prevent that, the literature implies, they will rejoice in a human-equivalent servant and companion. This article is of possible interest to those in the fields of the anthropology of work, anthropology and literature, futures research, and ethnic relations.

Morton Klass has taught anthropology at Barnard College, Columbia University, since 1965, and has been director of the Southern Asian Institute, Columbia University, since 1982. He received his B.A. from Brooklyn College in 1955 and his Ph.D. from Columbia in 1959, and has taught at Bennington College. His research interests and numerous publications pertain to South Asian society and culture, particularly stratification, culture change, anthropological theory, and change and modernization.

NOTE: An earlier version of this article was read at the 1982 meetings of the American Anthropological Association, Washington, D.C., and was published in the journal *Cultural Futures Research,* 7 (1983). Reprinted by permission.

THE robot—the manufactured equivalent of the human—is almost here. There are those in fact who would claim that it has already arrived and that quite a few people are currently engaged in the process of adjusting to its presence. There is another argument, however, one reflected in this article, that in an important sense manufactured equivalents of human have been with us for most of this century if not longer and that there is a substantial body of literature dealing, speculatively and varyingly but still significantly, with the implications, opportunities, and problems inherent in the association of humans and the manufactured equivalents of humans.

"Manufactured equivalents of humans" is admittedly an awkward way of putting it, but it does have its uses. I emphasize the word "equivalent" because the term introduces an important anthropological dimension, that of the alien—the person who in many societies is viewed as not of us, not truly human but only an equivalent of the true human. The views held in a given society of the nature of the alien may be quite complex. The alien may be simultaneously scorned and feared, for example, or considered both nonhuman—say, nonmarriageable—and yet very human—sexually quite desirable. It is certainly legitimate, therefore, for the anthropologist to inquire into the extent to which the manufactured equivalent of a human has reflected the particular perceptions of the alien that is characteristic of Europe-derived societies.

I also emphasize the word "manufactured" because it is the most satisfactory solution I could find for the rather knotty problem of defining what is meant by "robot." I intend it, however, only as a provisional solution, for the questions remain to be answered. Is the robot a machine only, and never anything else? Suppose it were constructed of living tissues, by genetic engineering? Suppose, whatever its composition, it could somehow reproduce itself? Or does the term imply the issue not so much of construction and composition but rather of human-like appearance? In other words, is it a robot because it has the equivalent of a human brain, or of human-like appendages, or because it moves or communicates like a human? Would it still be a robot if it were enclosed in a featureless box? Or is the term "robot" intended to reflect some sort of intellectual capacity? Might we call any programmed or programmable machine a robot or should we reserve the term for the machine that can program itself, that can learn, that can think? Or it is all of the above, or at least many, in some kind of combination?

Most of these questions, and many related ones, have been pursued in the voluminous literature of science fiction during the past half century. They become suddenly of anthropological interest, however, once we ask how our society has responded to the presence—if only fictional—of robots as aliens, as human-equivalents. The anthropologist will want to know the particular ways in which the robot is perceived as equivalent to humans, and the ways it is not. After all the alien is known to all human societies, but there is considerable variation from one to another about the supposed characteristics or attributes of the alien—that human/nonhuman. Is the robot, for example, marriageable or just sexually desirable? This is not a flippant question: we shall see that it is worth pursuing.

Some readers might observe with impatience that most of the foregoing questions become irrelevant if we but turn our attention to the actual contemporary industrial robot: unquestionably a machine, and a particular kind of machine, with specific capacities and limitations. I would argue, however, that it is precisely for this reason that a brief review of the fictional history of the robot becomes a useful exercise. From its first appearance, the term "robot" has undergone a series of transformations, though no living—nonfictional—human had ever actually encountered one of the entities encompassed by the term. It follows then that an analysis of these transformations will illuminate what people find attractive about robots—about even the notion of robots—and what they find disturbing or even frightening.

ČAPEK'S ROBOT AND SHELLEY'S CREATURE

The term was introduced to the English-speaking world with the production in 1922 of Karl Čapek's play, *R.U.R. (Rossum's Universal Robots),*[1] and is derived from the Slavic root for "work" or "worker." Let us observe, therefore, that in Čapek's play the robots are played by human actors; they are not machines of any kind, but rather protoplasmic beings that not only resemble but are supposed to resemble ordinary human beings. They differ from humans, however, in a number of ways—perhaps most significantly in that they are not born but are manufactured. Thus if we can call them machines at all we would have to say that they are machines that look, and to a great extent act, like humans.

Čapek's robot, in fact, was remarkably similar to Mary Shelley's creature in her famous novel, *Frankenstein, or the Modern Prometheus,* written approximately 100 years earlier, first published in 1817.[2] Why then do we not, at least retroactively, refer to Shelley's creature as a robot? Further, why did science fiction writers of mid-century—obviously fully aware of their debt to Čapek for the term—proceed to transform the robot into a sentient, vaguely humanoid, but nevertheless thoroughly mechanical machine? Still further, let us note that this remarkable change was fully endorsed by the readers of science fiction and ultimately by our society at large, that latter endorsement being reflected in the recent appearance among us of actual machines that are everywhere called robots.

I would argue that these two issues—the apparent reluctance to refer to Mary Shelley's creature as a robot, and the swift transformation of the robot in science fiction and popular usage from a flesh-and-blood being to a thing of metal, glass, and plastic—are very much related to each other. Examining them together provides us with important insights.

Mary Shelley's creature was, one might argue to begin with, actually the first android to appear in fiction, not the first robot. The term "android" came into use during the 1930s in science fiction to denote protoplasmic creations, of the kind Shelley and Čapek had described, or at least beings, whatever

1. *R.U.R. (Rossum's Universal Robots),* trans. Paul Silver (Garden City, NY: Doubleday, Page, 1923).

2. *Frankenstein, or the Modern Prometheus* (New York: Airmont, 1963).

their composition, capable of passing under superficial if not thorough examination as humans. The science fiction writer and editor Lester del Rey has in fact suggested that Čapek's "artificial men would now be called androids."[3] Nevertheless I would suggest that Čapek's "artificial man," though it looked like what was later to be called an android, was in important ways the first robot, the fictional forerunner of the machines that go by that name in modern factories. On the other hand I would have no objection to calling Mary Shelley's creature an android. There is a distinction here worth making, and it is interesting that it is made in the subtitles of the very two works in question.

THE NEO-FAUSTIAN QUESTION

Mary Shelley subtitled her book *The Modern Prometheus.* Her concern quite clearly was the tragedy inherent in the conflict between human aspiration and divine interdiction. There are echoes of Genesis in her work, and of the Faust legend: we are asked to contemplate the plight of the human who dares to do that which the gods have forbidden or at least reserved for themselves, and who thereby endangers his soul. On the other hand the key word in Čapek's subtitle—*Rossum's Universal Robots*—may well be universal. Why universal? Presumably because, though they are factory productions, these robots are versatile; they can perform any industrial task. Since in this regard they are like humans, it follows that they can, ultimately or potentially, replace humans. That becomes the central issue of the play, and therefore in this respect Čapek's robot exactly resembles the machine that goes by that name in the factory in Detroit, and the term is appropriate in both contexts.

Čapek's concern, let us observe, was not the Faustian question at all—that is, actions leading to an individual's damnation. His concern, rather, was what might be termed the neo-Faustian question: can a society as a whole aspire to the forbidden, can it set its collective foot on a path that will endanger its collective, or Durkheimian, soul? This is a recurrent question in the literature of this century, of course, and the path most commonly viewed with trepidation as the one leading to perdition is that of industrialization, even more specifically that of assembly line production. From this perspective *R.U.R.* takes its place alongside such works as Robert Graves's *Watch the North Wind Rise*—how we destroyed the machines and found God—and most particulary Aldous Huxley's *Brave New World.*[4] In Huxley's novel, for example, we knock the tops off our crosses to convert them into Ts—as in Henry Ford's Model T. Having thus symbolically damned ourselves, we serve the demonic assembly line by genetically engineering humans to fit it. In *R.U.R.* we do exactly the same thing, but by replacing humans with workers manufactured in the factory itself.

Science fiction writers and readers of the 1930s and 1940s, however, exhibited

3. *The World of Science Fiction: 1926-1976—The History of a Subculture* (New York: Garland, 1980), p. 87.

4. *Watch the North Wind Rise* (New York: Creative Age Press, 1949); the novel was originally published in England as *Seven Days in New Crete. Brave New World* (New York: Harper & Row, 1968) was first published in 1932.

little fear of technology. Rather it was the common belief of this genre that technology and science, or at least the practitioners thereof, had the capacity to solve our—then—current problems, cope with future ones, and thus ultimately usher in some form of Golden Age.[5] There is no single theme that encompasses all of science fiction, of course, but this confidence that humans could cope with the problems posed by science and technology was certainly pervasive in the literature, and so was an awareness of what I have called the neo-Faustian question. In the last of his robot stories, Isaac Asimov—one of the most well-known and influential writers of science fiction—has a character say,

> The Machine is only a tool after all, which can help humanity progress faster by taking some of the burdens . . . off its back. The task of the human brain remains what it has always been; that of discovering new data to be analyzed, and of devising new concepts to be tested . . . . These reactionaries . . . claim the Machine robs man of his soul. I notice that capable men are still at a premium in our society; we still need the man who is intelligent enough to think of the proper questions to ask.[6]

Indeed this entire series of stories by Asimov, originally published between 1940 and 1950, may in part be interpreted as a conscious and direct confrontation with the dilemma underlying the neo-Faustian question, Will the robot—the sentient universal machine—ultimately destroy us, or destroy everything that makes life worth living? Or will we be able to remain in control, keeping the robot forever in subjection, as our servant? Asimov, like many other science fiction writers of the period, was obviously aware of the supposed threat that was, or would be, posed by the appearance of robots among us, but in his work he argued that humans would find a way to retain their dominance over the machine. The symbol of his solution, and of his confidence in continuing human hegemony, was his set of laws, the "Three Laws of Robotics" to be implanted in every robot at the time of manufacture:

> 1. — A robot may not injure a human being, or, through inaction, allow a human being to come to harm.
> 2. — A robot must obey the orders given it by human beings except where such orders would conflict with the First Law.
> 3. — A robot must protect its own existence as long as such protection does not conflict with the First or Second Law.[7]

THE ROBOT AS SERVANT

The robot, in other words, must become and must remain the servant of the human. This theme—the robot as permanent and perpetual servant of humans, despite all improvements in the manufacture of robots and all declines in human capacities—is expressed again and again in the science fiction of the

5. In his study of an early and influential science fiction organization, Damon Knight, a leading science fiction writer and editor, quotes Donald A. Wollheim, another well-known editor, on the views of some of the science fiction aficionados of the mid-1930s: "[Will] Sykora believed that science-fiction was going to lead somehow-or-other to some great-I-don't-know-what. And he had a good precedent for it, because [Hugo] Gernsback [the first science fiction editor] was writing the same kind of thing—how science could save the world, and so on. And I remember the time [Forest J.] Ackerman coined a slogan, 'Save Humanity with Science and Sanity.'" Knight, *The Futurians* (New York: John Day, 1977), pp. 13-14.

6. From a story first published in 1950, reprinted in *I, Robot* (New York: Fawcett Crest Books, 1970), pp. 187-88.

7. Ibid, p. 6.

middle of the century. In Clifford Simak's celebrated *City* series (1952),[8] one robot servant, a sort of mechanical Jeeves, serves humans for thousands of years, at times humbly, and at times as nurse, teacher, or even keeper, but always intrinsically as a servant. In other words, one response to the neo-Faustian dilemma that could be found in many science fiction works at mid-century was, in effect, "Yes, the robot is a potential threat, and a serious one, but humans are masters of their fate and will engineer the robot so that it will never get out of control. The human will remain the robot's master forever, and the robot will remain the eternal servant."[9]

Some might argue that this is in its way as theological a position as the neo-Faustian question that provoked it. Theological issues to one side, there is of course a fundamental problem here, carefully avoided by Asimov but expressed by a character in a recent novel by another major writer of science fiction, Robert A. Heinlein:

> When I was a student, I read some classic stories about humanoid robots. They were charming stories and many of them hinged on something called the laws of robotics, the key notion of which was that the robots had built into them an operational rule that kept them from harming human beings either directly or through inaction. It was a wonderful basis for fiction . . . but in practice, how could you do it? What can make a self-aware, nonhuman, intelligent organism—electronic or organic—loyal to human beings?[10]

Has the wheel then come full circle in 50 years? Are we not hearing in Heinlein echoes of Čapek? The robot is, or can be, the servant of the human, but if we once suspend as irrelevant our pious assumption of eternal human hegemony do we not see a danger lurking beneath the metal surface of our servant? Suppose one day it becomes abruptly no longer "loyal to human beings." After all, as Heinlein says, why should it be?

One reason for its loyalty, perhaps, is that we humans seem to have developed an affection for the robot. It is the affection of a superordinate for a permanent underling, perhaps, or the affection we would show to a pet. Indeed the roles permitted the robot in science fiction are remarkably restricted: the robot may be a servant, as we have seen, but apparently never an employee. And, although we see in the robot a potential threat to our well-being, that threat is almost never a sexual one, despite the fact that sexual threat does customarily characterize the alien in Europe-derived societies.

I suggest, in fact, that this last omission—the absence of any perception of the robot as a sexual threat—played an important part in the transformation of the robot in literature from an infernal danger into something to which we respond with pleasure and even affection.

ROBOT VERSUS ALIEN

Is the issue a ludicrous one? After all, one might argue, a machine—however

8. *City* (New York: Gnome Press, 1952), a collection of stories first published from 1944 through 1951.

9. Harry Bates examined some implications of this curious assumption in his classic "Farewell to the Master," *Astounding Science Fiction,* Oct. 1940.

10. *Friday* (New York: Holt, Rinehart and Winston, 1982), p. 96.

sentient—normally lacks the fundamental equipment for the sexual encounter. It cannot rape or be raped, it cannot even entice or beguile. While this is true, and of course contributes to the view of the robot as asexual or even nonsexual, it is also true that such deficiencies do not in other cases necessarily impede the ascription of sexual threat to aliens in Western, or Europe-derived, societies, particularly in popular legend and literature.

The dragon St. George slew wanted only virgin and nubile females, so its appetite could not have been solely gustatory, and the same thing must be said of the sea monster encountered by Perseus when he rescued Andromeda. If we cannot know for certain what King Kong intended to do with Fay Wray when he had a moment free from the attentions of giant snakes and biplanes, still we are in no doubt that his intentions were as dishonorable as those of the bull that kidnapped Europé.

The alien male may be a monster, but that does not prevent him from wanting to rape our women—it merely makes his desires more awful. Still "our" males are supposed to be sexually attracted to alien females. So, anyway, argues Herodotus, variously considered the Father of History, of Anthropology, and of Lies. He opens his history of the Persian Wars with the suggestion that the continual conflicts between the Greeks and the peoples of Asia Minor could be attributed to the kidnapping and raping of each other's women: "According to the Persians . . . the Phoenicians began the quarrel," he notes,[11] but he goes on to record that men of both sides joined in the game enthusiastically.

This Herodotean conviction—most particularly that aliens will find our women sexually attractive and will attempt to carry them off—apparently remains very much with us; it is certainly to be found in science fiction, at least on the more lurid magazine covers of mid-century. Again and again we are given a representation of a tentacled bug-eyed monster carrying off a struggling, half-naked, human female, while Perseus-of-the-future, raygun in hand, comes dauntlessly to her rescue. There were indeed covers on which rogue robots engaged in the same practice—for which, after all, they were at least as well equipped as the typical bug-eyed monster—but it would seem that such covers were much rarer. It is possible they simply did not sell as well. What I am proposing is that the robot—though often perceived as alien and dangerous—was nevertheless rarely perceived as constituting a sexual threat. This is significant, I would argue, because in Western society the alien is commonly seen as a source of both sexual threat and its converse, sexual attraction.

The robot as a sex object does occur in science fiction, but only rarely. An example of such an occurrence is Lester del Rey's story, "Helen O'Loy" (1938).[12] This is the story of a man who falls deeply in love with a female robot—that is, a sentient machine cunningly covered on the outside with plastic and rubber to convey the appearance of a female human. Though often referred to as a classic of science fiction, "Helen O'Loy"

11. *The History of Herodotus,* trans. George Rawlinson (New York: Lincoln MacVeagh/Dial Press, 1928), p. 1.

12. "Helen O'Loy," *Astounding Science Fiction,* Dec. 1938.

gave rise to no subgenre; it spawned no imitators.

Indeed even Mary Shelley's creature, lonely as it was, did not seek the arms of humans, and Čapek's robots may have killed humans and sought to replace them, but they never violated humans sexually, nor did they seek to make love to them. Perhaps then we may interpret del Rey's Helen O'Loy as still another robot servant, but one that provided a rather special service, one not of a kind that most people, or at least most writers, expect from robots.

It is perhaps worth observing, however, that Helen O'Loy, machine though she is, clearly reciprocates the love of her human husband.[13] From the time of Mary Shelley herself, writers have found themselves drawn to the minds and feelings of human-equivalents. Is such a creature lonely? Frightened? Resentful? Does it feel the alienation of being in a world it never made, but which made it? William Tenn, in a poignant story, "Down among the Dead Men" (1955),[14] explores in depth the feelings of bitterness such creations might be expected to harbor; he even suggests some of the contemptuous terms—"uties," "wombats"—they might use to express their resentment of those born in normal biological fashion. And if Asimov's "Laws of Robotics" reflect an effort to program human morality and ethics into the machines, some writers—such as Eando Binder in the tale of Adam Link (1939)[15]—suggest that if a machine were to be intelligent and self-aware, would it not also of itself be concerned with right and wrong?

CONCLUSION

The robot in science fiction then was portrayed at first as an alien and as a threat, but the danger was perceived as primarily an economic one—apart, that is, from the theological danger. The robot may drive us from our jobs and otherwise destroy our economic well-being, it was felt; it may even threaten to destroy the world as we know it; it may endanger our collective soul. But we have never believed it would dishonor or corrupt us, something we have always assumed that other aliens wanted most of all to do. Perhaps not surprisingly then we seem to be able to live with whatever threat, economic or theological, the robots represent; we do not exhibit horror or revulsion, or even very much trepidation.

Perhaps this is because science fiction writers and readers began to feel that, despite the potential dangers, robots actually held out a promise: not simply—if we are lucky and smart and

13. Though del Rey called her a robot, Helen O'Loy might well be termed an android, of course—a constructed creature that can pass as a human unless one accidentally peels back the rubber mask and reveals the gears and bolts beneath. Such a creature is indeed a fictional object of human sexual fears and desires. This theme is explored extensively in Heinlein's novel *Friday* (see note 10); and in a very different way in William Tenn's "Down among the Dead Men," *Galaxy Science Fiction,* June 1954, rpt. in *Of All Possible Worlds* (New York: Ballantine Books, 1955); and indeed in many other works. In a way then we might say that Čapek's robot had two offspring: the one we call robot and view as both servant and possible economic threat, and the one that came to be called android. Since the latter can pass as human and therefore move unnoticed among us, it can indeed easily become a sexual threat.

14. See note 13.

15. "I, Robot," *Amazing Stories,* Jan. 1939.

careful—to serve us all its days, but to be a companion to us in the universe, one more meaningful than any animal companion, someone to share and even assume our ethical burdens. The robot, I have noted, is never portrayed as an employee but rather as a servant—if we are fortunate, as a faithful servant. Perhaps such a companionship is particularly acceptable or desirable precisely because the sexual element is missing.

In any case, there is something that draws us to the robot, despite all our fears. We would like to embrace it—I say "it" advisedly—even though we do worry occasionally whether it will be necessary in the end to destroy it before it destroys us.

Report of the Board of Directors to the Members of the American Academy of Political and Social Science for the Year 1982

MEMBERSHIP

MEMBERSHIP AS OF DECEMBER 31

Year	*Number*
1972	21,963
1973	21,070
1974	19,473
1975	16,923
1976	15,516
1977	14,202
1978	12,816
1979	10,884
1980	10.059
1981	9,874
1982	9,536

FINANCES

SIZE OF SECURITIES PORTFOLIO

MARKET VALUE AS OF DECEMBER 31

1972	642,808
1973	533,024
1974	371,004
1975	440,450
1976	504,046
1977	451,545
1978	385,795
1979	377,915
1980	368,926
1981	351,886
1982	390,119

PUBLICATIONS

NUMBER OF VOLUMES OF *THE ANNALS* PRINTED (6 PER YEAR)

1972	143,360
1973	132,709
1974	120,397
1975	104,049
1976	101,789
1977	91,367
1978	85,605
1979	71,513
1980	65,153
1981	69,313
1982	74,211

NUMBER OF VOLUMES OF *THE ANNALS* SOLD (IN ADDITION TO MEMBERSHIPS AND SUBSCRIPTIONS)

1972	16,721
1973	12,430
1974	13,153
1975	13,034
1976	12,235
1977	6,296
1978	8,124
1979	5,907
1980	8,751
1981	5,884
1982	7,562

STATEMENT OF INCOME AND EXPENSE FOR THE YEAR ENDED DECEMBER 31, 1982

Income	
Royalty—Sage Publications	$120,000
Sales of review books	1,951
Royalties and reprint permissions	5,866
Annual meeting revenue	7,589
Donations	250
Miscellaneous	549
Total Income	136,205
Operating Expenses	
Salaries	74,532
Payroll taxes	7,760

Pension expense	16,986
Employee benefits	618
Annual meeting expense	15,429
Depreciation	2,328
Insurance	1,625
Miscellaneous	19,721
Postage	1,747
Repairs and maintenance	4,879
Supplies	1,880
Telephone	1,851
Utilities	5,524
Total Operating Expenses	154,880
Loss from Operations	(18,675)
Other Income (Expenses)	
Investment income (net)	30,923
Gain on sale of investments	14,392
Grant administration overhead	12,200
Total Other Income	57,515
Net Income	38,840
Retained Earnings — January 1	322,915
Retained Earnings — December 31	$361,755

Report of the Board of Directors

During 1982, the six volumes of THE ANNALS dealt with the following subjects:

January — *Government and Economic Performance,* edited by J. Rogers Hollingsworth, Professor, Department of History, University of Wisconsin, Madison.

March — *The Internationalization of the American Economy,* edited by J. Michael Finger, Senior Economist, World Bank Development Policy Staff, Washington, D.C.; and Thomas D. Willet, Horton Professor of Economics, Claremont Graduate School and Claremont McKenna College, Claremont, California.

May — *Young Children and Social Policy,* edited by William M. Bridgeland, Associate Professor, and Edward A. Duane, Professor, Department of Social Science, Michigan State University, East Lansing.

July — *The American Judiciary: Critical Issues,* edited by A. Leo Levin, Director, and Russell R. Wheeler, Assistant Director, The Federal Judicial Center, Washington, D.C.

September — *International Terrorism,* edited by Marvin E. Wolfgang, President, The American Academy of Political and Social Science, and Professor, Department of Sociology, University of Pennsylvania, Philadelphia.

November — *Middle and Late Life Transitions,* edited by Felix M. Berardo, Professor, Department of Sociology, University of Florida, Gainesville.

The publication program for 1983 includes the following volumes:

January — *Housing America,* edited by Wallace F. Smith, Professor, Department of Business Administration, University of California, Berkeley.

March — *Implementing Governmental Change,* edited by Charles E. Gilbert, Professor, Department of Political Science, Swarthmore College, Swarthmore, Pennsylvania.

May — *The Global Refugee Problem: U.S. and World Response,* edited by Gilburt D. Loescher, Assistant Professor, Department of Government and International Studies, University of Notre Dame; and John A. Scanlan, Assistant Director, Center for Civil and Human Rights, Notre Dame Law School, Notre Dame, Indiana.

July — *Health Care Policy in America,* edited by S. E. Berki, Professor and Chairman, Department of Medical Care Organization, School of Public Health, University of Michigan, Ann Arbor.

September — *Nuclear Armament and Disarmament,* edited by Marvin E. Wolfgang, President, The American Academy of Political and Social Science, and Professor, Department of Sociology, University of Pennsylvania, Philadelphia; and Robert H. Kupperman, Executive Director, Science and Technology Center for International Studies, Georgetown University, Washington, D.C.

November — *Robotics: Future Factories, Future Workers,* edited by Robert J. Miller, Professor, Department of Anthropology, University of Wisconsin, Madison.

During 1982, the Book Department published approximately 350 reviews. The majority of these were written by professors, but reviewers also included university presidents, members of private and university-sponsored organizations, government and public officials, and business professionals. Over 700 books were listed in the Other Books section.

Eighty requests were granted to reprint material from THE ANNALS. These went to professors and other authors for use in books in preparation and to nonprofit organizations for educational purposes.

MEETINGS

The eighty-fifth annual meeting, which was held on April 23-24, 1982, had as its subject *International Terrorism* and continued the tradition of our gatherings with respect to the diversity of organizations represented by delegates, the size of the audiences, and the interest displayed. Ten embassies sent official delegations, as did eight United Nations missions and sixteen states, cities, and agencies of the federal government. Delegates were also sent by 65 American and foreign universities and colleges and 37 international, civic, scientific, and commercial organizations. More than 300 persons attended one or more sessions.

The theme of the eighty-sixth annual meeting, held April 22 and 23, 1983, at the Bellevue Stratford Hotel, Philadelphia, was *Nuclear Armament and Disarmament*. The September 1983 volume of THE ANNALS contains the papers presented at the meeting.

OFFICERS AND STAFF

The Board reelected the following officers: Marvin E. Wolfgang, President; Richard D. Lambert, Vice-President; R. Jean Brownlee, Secretary. Elmer B. Staats was elected Treasurer, replacing Howard C. Petersen, who retired from his positions as Board Member (since 1957) and Treasurer (since 1967). The following Board Members were reelected for three-year terms: R. Jean Brownlee, Richard D. Lambert, and Elmer B. Staats. Lloyd Cutler, who resigned from the Board when he was appointed Legal Counsel to the White House during the last year of the Carter administration, has agreed to return to the Board. Mr. Cutler and Randall M. Whaley were elected to three-year terms.

Respectfully submitted,

THE BOARD OF DIRECTORS

Elmer B. Staats
Marvin E. Wolfgang
Lee Benson
A. Leon Higginbotham, Jr.
Richard D. Lambert
Rebecca Jean Brownlee
Covey T. Oliver
Thomas L. Hughes
Randall M. Whaley
Lloyd N. Cutler

Philadelphia, Pennsylvania
1 September 1983

Book Department

INTERNATIONAL RELATIONS AND POLITICS

FALK, RICHARD. *Human Rights and State Sovereignty*. Pp. x, 251. New York: Holmes & Meier, 1981. $28.50. Paperbound, $15.75.

This book is a collection of a number of previously published articles focusing on various aspects of human rights: the relation between political and economic rights and between types of political systems and civil rights norms, the connection between legal formalism and actual observance, and the influence of American foreign policy, and of the United Nations, on the promotion of rights.

Richard Falk argues that a purely formal definition of political rights often hides socioeconomic inequalities and other substantive shortcomings within systems. He criticizes the United States for promoting certain procedural rights so as to further its national interest. He shows how the interpretation of human rights varies from country to country depending on their accepted theoretical foundations—for example, statist, hegemonial, naturalist, supranational,aand populist. He accuses the United States of basing itself on hegemonialism rather than on the much superior supranationalism or populism.

One may accept Falk's assessment of the Kennedy-Johnson years as a period during which human rights were promoted, as well as his critique of Kissinger for being "openly scornful of introducing human-rights concerns into serious diplomacy." But one might question his fulminations against Moynihan—and other neoconservatives—who allegedly "used human rights as an ideological tool against the Third World," and his praise of Carter and Andrew Young, whose concerns with human rights caused them to support the United Nations strongly.

Falk's preference for the supranational approach to human rights produces a certain myopia, which is reflected in an almost unreasonable admiration for the work of the United Nations and other international agencies and—despite imprecations against formalism—of the genocide convention, and similar international agreements. There is a brief reference to Amnesty International, but no critique of the blindness of this organization to violations of the rights of Jews in various countries. There is no serious critique of the hypocrisy of the United Nations, which is blind to the violations of more than half of the countries in the world.

Much of the book is an apologia for Third World countries. Thus while Falk condemns the statist logic of Western countries—which may lead them to protectionist or interventionist behavior—he condones such behavior by Third World countries because there it is "anti-interventionary and anti-imperial," and hence a manifestation of "the pure morality of the states system."

Another area in which polemics dominate objectivity is Falk's approach to capitalism. Perhaps he is right when he argues—in a stimulating review of the Trilateralists' book, *Crisis of Democracy*—that while neither capitalism nor socialism is perfect, "it seems clear that only socialism has the capacity to deal with mass poverty in the short run." But he goes overboard when he asserts, without proof, that "the Soviet role in the Third World has, up to this point, been less widely detrimental to the pursuit of human rights than has the American role"; and when he fails to criticize the USSR's own formalism as evidenced in the charade of its signing of the Helsinki agreements—in fact, there is no discussion at all of these agreements.

Falk's biases affect his logic. Thus he is entitled to be critical of the Americans' cold war stridency in general, and the Truman Doctrine in particular, but he is not justified in considering these ipso facto as expressions of hostility toward human rights; nor is he fair when he accuses the AFL-CIO, and other so-called right-wing critics of Soviet violations of human rights, of merely wanting to stir up trouble. One final point: Falk's romantic faith in Khomeini and his revolution, to which he devotes an entire chapter, has unfortunately not been justified by events; but even the violations that had been committed before the book was written are excused by the history of the Shah's cruelties and the exigencies of postmonarchical "revolutionary justice."

WILLIAM SAFRAN

University of Colorado
Boulder

KANIN, DAVID B. *A Political History of the Olympic Games*. Pp. 161. Boulder, CO: Westview Press, 1981. $17.00.

There are few today who would argue with Kanin's basic premises that the Olympic Games are as much a political event as a sports spectacular, and that the politicization of the Olympics is not of recent origin, but has been an integral part of Olympic history, ancient and modern. Olympic victories in the sixth and seventh centuries B.C. celebrated the political and military prowess of Greek city-states in the same fashion that success in the modern games is directly related to international prestige and ideological superiority for nation-states. In fact Baron de Coubertin's mottoes of fair play and the value of a participatory experience represent an ideological smokescreen for his covert purposes of revitalizing the Fench military and the moral integrity that had been devastated by the Franco-Prussian War. Only the most ideologically pure would deny that the Olympics have become an international political arena.

The political value of the Olympics seems to rest with its high visibility and its salience to general populations. Sport, particularly the international variety, conveys a country's capabilities to the general public like no other event, whether it be cultural or diplomatic. Thus for Kanin the Olympics are an exercise in public diplomacy with the avowed goals of demonstrating to the external world and the home constituency that the government and its supporting ideological systems are viable, active, and productive.

The first few chapters of *A Political History of the Olympic Games* review the ancient and modern games and describe the political events that occurred at London (1908 and 1948), Berlin (1936), Mexico (1968), Munich (1972), and so forth. Little of this discussion is significantly new, analytical, or insightful. Perhaps the most important point Kanin makes with respect to this descriptive material is that the Olympics, and international sporting competition in general, are serious but diplomatically safe. Countries can test tactics in foreign policy—for example, ping-pong diplomacy—without dire consequences. The same cannot be said for experimentation in legal, economic, and military strategy. Kanin is correct in his assertion that sport is actually quite peripheral to mainstream, hard-line international politics.

The diplomatic failure of the 1980 U.S. boycott of the Olympics can be viewed as resulting from the fact that the Carter administration did not have a realistic grasp on the role of sport in international politics. The chapter on the Moscow Olympics represents the most significant contribution of this book. Kanin, an employee of the Central Intelligence Agency, performed as a consultant to various agencies and persons who were formulating the boycott strategy. His description of the events leading to the boycott and his accounts of the political maneuvering that took place within European, African, and other nations affected is perhaps the best discussion available. For example, the Carter administration misplayed its hand when it overemphasized the political value of the Olympics and underestimated the political impact of the Soviet's treatment of dissidents. In addition the resistance to the U.S. proposal for a boycott pointed to a greater regionalism among nations and also to a reduction in the tendency to perceive world affairs as a function of bilateral superpower strategy. Kanin has demonstrated well the lessons that can be learned from a diplomatic rather than athletic study of international sport.

JAMES H. FREY

University of Nevada
Las Vegas

MURRAY, DOUGLAS J. and PAUL R. VIOTTI, eds. *The Defense Policies of Nations: A Comparative Study*. Pp. xvi, 525. Baltimore: Johns Hopkins University Press, 1982. $12.95.

This is an excellent book for those interested in the substance and process of defense policymaking. It complements two earlier volumes, *American Defense Policy* and *Comparative Defense Policy,* which, like the book under review, were edited by members of the Department of Political Science at the U.S. Air Force Academy. *Comparative Defense Policy* focused on four issues—military doctrine, force posture, the decision-making process, and weapons acquisition. In contrast *The Defense Policies of Nations* focuses on specific countries and how defense policy is made in these states.

Each state's defense policy is analyzed according to a common framework, developed by Murray and Viotti, consisting of (1) the international environment as perceived by the state; (2) the particular national objectives, strategy, and military force employment doctrine; (3) the state's defense policymaking process; and (4) major defense issues facing the state. By and large the contributors stick to this framework, which enhances the comparability of the individual studies. The country studies include analyses of the United States, the Soviet Union, China, the Western European states, Japan, and two smaller countries—Sweden and Romania. The bibliographic essays on each of the countries are very helpful.

Particularly valuable are the introduction and the conclusion, in which the editors assess the place of comparative defense policy within the context of international relations theory. While Murray and Viotti do not attempt to construct a theory, they do provide a theoretically oriented framework on which the defense policies of various states may be compared.

Despite the overall value of the book, there are several shortcomings. First, in several cases Murray and Viotti fail to indicate the sources of previously published material, such as chapters by Kenneth Waltz, Stanley Hoffmann, Michael Howard, Richard Pipes, Benjamin Lambeth, and Arthur Alexander. While some of the chapters have not previously been published, others are reprinted from widely read sources. Second, the print, particularly for offset quotations, is small and difficult to read. These are relatively minor weaknesses and do not detract from the intellectual value of the book.

The Defense Policies of Nations is an appropriate text for upper-division undergraduate and graduate courses in comparative defense and foreign policy. Those conducting research in this field will also find this book a valuable resource.

DAN CALDWELL

Brown University
Providence
Rhode Island

PERKINS, JOHN A. *The Prudent Peace: Law as Foreign Policy*. Pp. xvi, 246. Chicago: University of Chicago Press, 1981. $28.00.

In *Prudent Peace,* Perkins has attempted to explain the logic and necessity of resorting to law as a means of fashioning foreign policy. His central perception of the role of law as prudent foreign policy constitutes the basis for his belief that the prevailing attitudes about law have been focused on the wrong questions. He clearly perceives, to his credit, that "nations twist law to their own ends," that law is perceived as insufficient to achieve the goals nations set for themselves. Whether self-determination and nonintervention are principles of law or merely policy goals is an important question, and Perkins does not shirk his obligation to answer it. The book presents a concise discussion and analysis of the major legal issues between nations; unfortunately the solutions and the assumptions made to support them are often murky in thought and language.

Prudent Peace was conceived, according to Perkins, at the conclusion of World War II. It was executed during what he calls the "official calm of detente." But aside from the fact that detente was never official, and arguably never calm, we are now in the post-Brezhnev era. The dubious role played by law in the past is clearly subservient to the importance today of economics and personalities. The impotence of the United Nations and the imminent disintegration of the Organization of Petroleum Exporting Countries refute Perkins's contention that legal entities may bring order where there is chaos.

Perkins concedes that power rather than law shapes policy. "A pervasive cynicism reigns," he states; the world "has lost confidence" that any nation is prepared to respect law in matters it regards as vital. Nowhere is it demonstrated that the world ever held confidence that nations would respect law, absent the threat of force. Frequent reference is made to the postwar years. If we are correct in assuming that he means post-World War II years, how long do we exist in this era? Certainly the Korean and Vietnamese "conflicts" argue in favor of ending the "postwar" era. These conflicts also represent lessons different from those of World War II. To propose that we adhere to postwar lessons, Perkins must convince us that these lessons still apply, and this he does not do.

The problem of occupied lands—Gaza and the West Bank—and divided countries—Cyprus and others—is addressed, but answers are not forthcoming. Having spent the turbulent years of 1973-74 in the Middle East, most of it on Cyprus, I was interested in how Perkins would perceive the problems that have existed there for eons. Scholarly intuition, unfortunately, is no substitute for actual experience.

Perhaps the scope of the book is too broad; perhaps nations cannot be made to adhere to law, absent force. Perhaps Perkins should provide concrete answers to the "how" of compliance, including enforcement of his "Declaration of Policy," rather than esoteric explanations of "why." Ultimately *Prudent Peace* represents a modest addition to the universe of information on international law; it may provide the springboard for great works yet to be conceived.

JOSEPH E. KALET

Bureau of National Affairs
Washington, D.C.

TILL, GEOFFREY et al. *Maritime Strategy and the Nuclear Age.* Pp. x, 274, New York: St. Martin's Press, 1982. $35.00.

This excellent book on naval strategy was written by Geoffrey Till, a member of the history department of the Royal Naval College, Greenwich, and a lecturer at the City University, London. He is the author of *Air Power and the Royal Navy* (1979). Associated with him are seven collaborators. One, John Hattendorf, is professor of strategy at the Naval War College, Newport, Rhode Island, and another is Craig Symonds, a professor of history at the U.S. Naval Academy, Annapolis. The others are English scholars who have published widely in the field of naval history and maritime affairs.

This is a well-organized and scholarly book that had much to say to a landlubber like me. It includes a brief review of the influence of sea power on history, and spells out the main constituents of maritime strategy. A valuable chapter succinctly describes the literature on the subject, from the age of the galley to the nuclear age, with appropriate coverage devoted to Alfred Thayer Mahan and the spokesman for Russian thinking on the subject, Admiral S. G. Gorshkov.

A navy does not, of course, exist in a vacuum, and accordingly a third chapter describes the main elements of naval power, the nature of the maritime community, and its resources, style of government, geography, shipping, and bases. The concept of the decisive battle, which is taken with a grain of salt, is then reviewed. Successive chapters deal with the concept of command of the sea and the exercise of command, with special reference to the problem of maritime communications.

The new environment for navies in the contemporary world is then analyzed, particularly in connection with the new technology. Today navies are utilized for sea control, the defense of sea lines of communication, the projection of power ashore, naval diplomacy, and strategic deterrence.

How extensive and important naval operations still are is spelled out in a closing chapter that reviews in detail all the naval operations carried on in the world in 1979, as well as the naval aspects of the Arab-Israeli War of 1973. Of course the airplane and nuclear weaponry have made a difference, but navies still play an important, often decisive, role in the military affairs of nations, large and small, in our day.

Till concludes that "neither sea power nor the ability to command the sea are absolute, but merely relative questions of degree," that "worthwhile parallels can still be drawn between past and present maritime experience," and that the range and extent of recent maritime events "confirm rather than deny traditional ideas about the importance of the sea to the security and prosperity of most countries."

Though the subject is necessarily technical, this is a well-written book well within the range of the general reader who is willing to make the effort, and worthy of the attention of political scientists.

DAVID FELLMAN

University of Wisconsin
Madison

AFRICA, ASIA, AND LATIN AMERICA

ALEXANDER, ROBERT J. *Rómulo Betancourt and the Transformation of Venezuela.* Pp. viii, 737. New Brunswick, NJ: Transaction Books, 1982. $39.95.

Robert J. Alexander's *Rómulo Betancourt* is a story of the transformation of Venezuela told through the biography of its foremost politician. In the 676 pages of narrative and notes, Alexander portrays Rómulo as part idealist and part practical politician. This winning combination helped the man to analyze his country's problems early; to propose solutions in the form of a program; and to compromise, when necessary, to bring it into effect. In abundant detail, which at times borders on the irrelevant, Alexander chronicles the struggles to form a party, the years of exile and underground activities, and the periods when Rómulo held high national office. The book successfully conveys a feeling for the boundless energy and dedication that culminated in his becoming the first democratically elected Venezuelan president to serve a full term, thus establishing a tradition for *Venedemocracia.*

Despite the rich detail that evidences years of painstaking research and the chronological completeness of the biography, Alexander fails to provide us with enough comparisons between Betancourt and other Latin American political leaders to give us an adequate perspective on the man's significance. This is all the more tragic if we recall that Alexander has published some two dozen books, which include biographies of Arturo Alessandri of Chile and Juan Domingo Peron of Argentina. Adjectives such as "mystical," "popular," "macho," and "magical" make Rómulo sound like a temporally misplaced or updated version of the nineteenth-century stereotype of the mystical *caudillo.* He also seems to fit, in some respects, the description of a populist, but nowhere is he called one. Perhaps more comparisons and some labeling—however facile or qualified—might have helped the general reader see beyond the trees to remember and appreciate the forest.

SUSAN RAMIREZ

DePaul University
Chicago
Illinois

HUGHES, COLIN A. *Race and Politics in the Bahamas.* Pp. vi, 250. New York: St. Martin's Press, 1982. No price.

This is a finely crafted small-scale study of the origins of black political consciousness in the crown colony of the Bahamas since World War II. It stresses the mechanics of parliamentary, democratic processes by which a growing self-identity was translated into party allegiances and ultimately into political triumph during and after independence. At the same time it demonstrates how this kind of limited-focus approach can enrich our understanding of the interrelationships among such variables as race, class, party politics, and ethnic consciousness in the rapidly developing societies of the Caribbean. I can recommend it unreservedly for its clarity of vision, its unpretentious style, and its insight into the functioning of a modern electoral system, Third World style.

I say "limited focus" not as a criticism, but because it is Hughes's considered intention to zero in narrowly on political processes and political personalities. He is superbly qualified for this task, with long-standing personal knowledge of events and actors; he has the necessary experience in Bahamian politics and knows well its intricacies and vicissitudes. Thus in introducing the book he abjures a wider theoretical goal, while nevertheless making excellent use of concepts and definitions from the abundant sociological and anthropological literature. So while the scope is limited, the study is well balanced by the use of comparative materials; and it is always informed by a sensitvity to the grander issues.

What points does Hughes make? First, he shows how black consciousness arose in an autochthonous form in the islands, responding to unique historical conditions and precedents. This is important because the U.S. Black Power movement bulks large in the background and could be mistaken as a prototype or model. Hughes shows how a meticulous analysis of parliamentary debates, newspaper reports, speeches, and so on can clarify local issues and their relationship to spontaneous political development at the national level. What Hughes shows is not simply how a black political awareness arose in the course of independence, but how a more fully focused ethnic identity developed which incorporated both nationalist and racial themes. In this context race and class are well distinguished and the socioeconomic issues clarified. Also well presented are the rise of a local black bourgeoisie and the formation of an emergent Bahamian nationalism in the face of pan-Caribbean currents.

What Hughes sets out to do, he does very well indeed. Still there are a few flaws—mainly of omission. First, one wishes that he had provided more background on the sociological structure of the island and on class structure—black and white. The class composition of the various parties and districts is only vaguely described, and one is left with the impression that most whites belonged to the Bay Street elite. Despite these minor criticisms, I found the book useful reading; any student of the Caribbean will do likewise.

DAVID D. GILMORE

State University of New York
Stony Brook

JUERGENSMEYER, MARK. *Religion as Social Vision: The Movement Against Untouchability in Twentieth Century Punjab*. Pp. x, 354. Berkeley: University of California Press, 1982. $30.00.

Nowhere more characteristically than in South Asia do people create visions of moral order and agendas for political action in religious terms. Yet for most Western and many Indian minds alike, to chart the road to a better India means to abandon religious for secular thought. The assumption that secular philosophies hold the key to India's future, however, stumbles over the simple fact that the everyday folk in that future envision their lives, their society, and moral action as inherently religious. In this richly descriptive book about Untouchable activism in India's Punjab state, Juergensmeyer shows how social reform politicians work among their poor fellows to better society as a whole, using religion to assault the religion-based inequities of caste.

Laid out in 24 short chapters, based on interviews, on field work, and on surveys, his story centers on the Ad Dharm movement, its leadership—above all Mangoo Ram—and its organizational career, from the 1920s to the 1970s. Narrative, rhetorical, and biographical detail bring the main characters to life. Juergensmeyer's affection for his principals and their cause gives the whole book a warm glow.

Juergensmeyer's argument is simply that social change, like stasis, can be understood and sought in the Indian context in religious terms. Moreover, he continues, there may indeed be no other meaningful ideological medium for politics for the Punjabi masses. In this respect Untouchables would seem the limiting case. As the most disenfranchised among Hindus, however, even they mobilize to change their society as a political sect. Secular communists and Congress social democrats are for his characters alien political animals.

Juergensmeyer is interested in narration, not explanation. How much of what he argues applies to Untouchables outside the Punjab, or to low castes in states where secular politics has sounder footing? Do movements like the Ad Dharm—small, fragmented, and ephemeral—actually accomplish anything? The social reforms Ad Dharmis sought were in fact put into effect by secular government action. Do such movements, to provide even the illusion of effectiveness, rely upon some political structure ideologically removed from competing, religiously motivated groups? By looking at this movement from the inside and by identifying with its leading personalities, Juergensmeyer asserts their position with force and clarity. But like them he seems to be going nowhere at the book's end.

DAVID LUDDEN

University of Pennsylvania
Philadelphia

JUPP, JAMES. *Party Politics: Australia 1966-81*. Pp. viii, 232. Boston: Allen & Unwin, 1982. $28.50.

Since the mid-1960s the frozen Western party systems generally have faced depression, social changes, new issues, and voter volatility. Jupp's study reviews the Australian version of these experiences since 1966. His starting point is logical in relation both to Western political trends and to Australian domestic politics, where it coincides with the retirement of the "two symbolic leaders of the 'old Australia'"—Sir Robert Menzies and Arthur Calwell. The book examines in separate chapters the organization, actions in government, and policies and ideologies of the three major parties—Liberal and National Country, together representing the main modes of conservatism, and the Australian Labor Party (ALP), embodying a complex mixture of radical, reformist, and Catholic traditions. The symmetry of treatment for each party assists the drawing of interparty comparisons.

While there are clear undertones of old orthodoxies regarding the differences between the parties—their links with organized interests and crude middle-class/working-class dichotomy—the study is valuable above all for its portrait of the contemporary condition of the parties, the blurring and overlapping of the social, ideological, and programmatic boundaries between them. The two largest parties, Liberal and Labor, are each revealed as a mixture of tendencies, partly due to the nature of their respective political traditions, but greatly enhanced by the events and climate of Australian politics since 1966.

Jupp's perspective is that of a Scandinavian-style social democrat. His study is neither shaped to a rigorous theoretical framework nor heavily burdened with theoretical concepts. If the idea of hegemony slips in occasionally, no wonder. For in a telling footnote statistic we learn that of 43 state and federal elections, 1964-80, the Labor Party has won only 14. Jupp argues that certain consequences have flowed from this and that Australia is as it is partly because of the political dominance of the conservative parties. That in turn is partly a result of the nature and deficiencies of the ALP.

The book is jargon-free and readable. There is some unevenness, stemming from the inclusion of both state and federal parties, and some overlapping of material in different sections. The worst passages read as stark summaries of events; the best are succinct and illuminating. Explanatory footnotes are generously used, but one more—to explain the term "preferences" for foreign readers unfamiliar with the electoral system—would have been useful. When tracing sources from the footnotes, the thematic organization of the bibliography is a constant irritation.

E. P. AIMER

University of Auckland
New Zealand

KHÁNH, HUỲNH KIM. *Vietnamese Communism 1925-1945*. Pp. 379. Ithaca, NY: Cornell University Press, 1982. $25.00.

This impressive book, based on an exhaustive investigation of French and Vietnamese primary source materials, will in future be required reading for anyone seeking to understand this complex period of Vietnamese history. Khánh has two aims. First, he chronicles the internal developments of the Indochinese Communist Party, including discussions of its leadership, organization, ideological orientations, strategies, and tactics, and its relationship with both international communism and Vietnamese society. Second, he seeks to assess the reasons for communism's success in 1945.

Khánh's book differs from a number of recent studies covering the same period, such as those of Duiker, Marr, and Pike. These other authors initially became interested in Vietnamese communism during their tenure with the U.S. government or military as part of their efforts to understand the success of the communists in the face of seemingly insurmountable odds. Khánh writes more as a Vietnamese patriot, concerned with tracing the recent history of Vietnamese radicalism, to tell "the story of a few thousand young men and women who challenged conventional wisdom, questioned the *raison d'être* of the centuries-old social and political order, and dedicated their lives to regaining Vietnamese independence." From this perspective he is critical of those, like Duiker, who take a very broad view of nationalism and include any Vietnamese who expressed some political opinions as "nationalists." Khánh in contrast argues persuasively for two major disaggregations in Vietnamese politics: the first, in the mid-1920s, which divided Franco-Vietnamese collaborationists from revolutionary patriots; and the second, from 1928-30, which divided revolutionary patriots from proletarian internationalists.

Khánh sees the history of Vietnamese communism as the union between indigenous Vietnamese and imported Marxist-Leninist proletarian internationalism, and the book traces the ebb and flow of this fusion. His analysis of the paradoxical influence of international communism on the indigenous movement is fascinating. He contrasts the failures of the movement in the 1930s to the successes of the 1920s and 1940s. Failure occurs when the Vietnamese communists, under the strong influence of the Comintern, adopt an antinationalist stance and abandon national independence as a goal; success comes from an emphasis on patriotism and national liberation. Although the importance of the international dimension of Vietnamese communism "cannot be exaggerated," Khánh makes it clear that the local party's popularity and successes came during those periods when the Vietnamese communist movement was isolated from its internationalist network. Most especially, in contrast to developments in Europe, Vietnam's August (1945) Revolution and power seizure ocurred without any external communist assistance.

STANLEY ROSEN

University of Southern California
Los Angeles

EUROPE

BROWNING, REED. *Political and Constitutional Ideas of the Court Whigs*. Pp. xi, 281. Baton Rouge: Louisiana State University Press, 1982. $27.50.

No political scientist or lover of eighteenth-century England should neglect this book. Essentially it is a group of articles on the careers and political thought of five influential supporters of the ministries of Sir Robert Walpole and the Pelham brothers, 1721-56—Lord Hervey, the courtier; Benjamin Hoadly, the controversialist; Thomas Herring, the preacher; Samuel Squire, the historian; and Lord Hardwicke, the lawyer. These are supplemented by a general article on the political thought expressed in the pamphlets supporting the policies of Walpole and the Pelhams.

Since the views of all these preachers and pamphleteers were very strongly influenced by their knowledge of the fall of the Roman republic in the first century B.C., the articles are preceded by a discussion of Cato the Younger's career as perceived by both Whigs and Tories, and followed by an account of the ever-increasing influence of Cicero, then universally called "Tully," on eighteenth-century British minds. Among the biographical articles, the one on Samuel Squire strikes a somewhat divergent note. It is not chiefly focused on Court Whig thought about current affairs, but on the early eighteenth-century desire to trace the roots of English parliamentary government into the Anglo-Saxon and Teutonic past. Since Squire not only did this, but later revised his views after further study of the Norman conquest and the reign of Edward I, this article should be of great interest to medievalists and teachers of English and American constitutional history.

Though Reed Browning's work will be a delight to readers with sufficient knowledge of British and Roman history to appreciate it, some sacrifice of detail, especially in the article on Squire, would have made possible the inclusion of background material needed by the general, especially the student, reader. Even the specialist in political thought may need to be reminded of Britain's then small population, of the still unreformed political system, especially in Scotland, and above all of how select an oligarchy the governing class of land-owning nobility and gentry really were. Also it seems a bit odd to devote a book to the political ideas of the Court Whigs without saying more about the first two Hanoverian kings and life at court during their reigns. George I is absent, and George II and his son Frederick are hardly ever mentioned.

HOLDEN FURBER

University of Pennsylvania
Philadelphia

EDELSTEIN, ALAN. *An Unacknowledged Harmony: Philo-Semitism and the Survival of European Jewry*. Pp. xii, 235. Westport, CT: Greenwood Press, 1982. $29.95.

It must be said at the outset that this is not an attempt to minimize the degree to which Jews have been the object of discrimination and persecution throughout Europe, during the Christian era up to this century. Mr. Edelstein's study is not, as the dust jacket proclaims, "a radical revision of Jewish history" and would fail in intellectual integrity if it claimed to be so. It is at most a corrective view emphasizing the role of ideologies and individuals who stood, with some success, against the evil that threatened Jews consistently and came near to triumph several times. Edelstein presents as his thesis this question: why was it possible for the Jews to survive in the face of consistent and virulent persecution replete with massacres, expulsions, and forced conversions? He credits several religious, cultural, and political currents, but above all the quality of philo-Semitism, as opposed to anti-Semitism, which spokesmen for these currents and numerous individuals were said to possess.

1. There were theological forces for philo-Semitism, first within the medieval Roman Catholic Church, later among Protestant denominations. Theological causes for toleration, at least for protection from physical destruction, were frequently but not quite consistently enunciated by a succession of popes. As the older faith from which Christianity sprang, as the people of the Savior, and in accordance with the Pauline doctrine looking to the—voluntary—conversion of the Jews, they were to be preserved.

2. Economic utility of the Jews to others, which is often accorded a major role in explaining Jewish survival, is granted but somewhat minimized by Edelstein, who demonstrates, correctly in my view, that an economic role was as often played by non-Jews, and moreover caused competition as often as acceptance. The economic motive is granted, however, concerning medieval Poland, to which Jews were called to fill just such a role: it created for some centuries a bright interlude in Jewish life.

3. Humanism, evolving into classic liberalism, the thought of the Enlightenment and of the French Revolution, with its proclamation of the rights of man and eventual extension of them to Jews, led most power-

fully toward emancipation, the course of which is sketched for several major countries and regions of Europe.

4. Modern nationalism in its early liberal manifestation expanded emancipation but at the price of demanding an end to Jewish cultural—not religious—separatism, a price Western European Jews gladly paid. Finally, late nineteenth-century nationalism turning exclusivist and xenophobic in Central and Eastern Europe, assimilated pseudoscientific racism, and along with it laid foundations for the extreme anti-Semitism that culminated in the Nazi movement.

Edelstein is careful to stipulate that philo-Semitism demands more than an attitude incidental to one's main purposes, be they Christian theology, economic betterment, or Enlightenment and liberal traditions and the achievements of their political goals. Was the motive of liberals, for example, primarily philo-Semitism, that is, a special concern for Jews, or were they simply following the inherent logic of the ideologies they professed and the policies they pursued? A few incidental observations may be in order.

1. This study rests almost entirely upon secondary sources, including, it must be said, many of the most distinguished historians of Jewish and relevant European experience. To add substance to Edelstein's thesis, greater use of primary sources such as memoirs and documents would be desirable.

2. It is difficult to prove the existence of a sentiment, essentially a psychological dimension, in retrospect except through memoirs and documents, and when these are cited as it were second hand, it does not add up to overwhelming empirical evidence.

3. An aspect of Enlightenment thought, barely touched on, is virulent anti-clericalism with drastic consequences during the French Revolution. Such an attitude was conducive to philo-Semitism, perhaps in a spirit of spite for the Christian clergy.

4. Edelstein's method is troublesome. As he identifies the major forces of philo-Semitism previously summarized, he attempts to distinguish limitations to his own argument, excluding self-seeking aspects of such thought and thereafer asserting the presence of philo-Semitism, with scant remaining proof.

The dismal view of Jewish history and its important modern corollary of Zionist ideology are not negated by this study, nor does it give great reassurance to Jews. The record of anti-Semitism and persecution is too overwhelming for that.

HANS E. SEGAL

Cleveland State University
Cleveland
Ohio

FIDLER, JOHN. *The British Business Elite, Its Attitudes to Class, Status and Power.* Pp. 384. London: Routledge & Kegan Paul. 1981. $40.00.

Fidler had agreeable access for first-hand interviews to 130 topflight chairmen, managing directors, and other directors from a range of large British companies. They were "members of the elite within the business world, men of genuine wealth and power." They included "four men who at some stage have headed nationalized industries, a former Lord Mayor of London, two ex-MPs, two presidents of the CBI [Confederation of British Industry] and several members of its Grand Council, sixteen known millionaires and five of the heads of the ten largest industrial companies ranked by turnover."

Those interviewed headed manufacturing, finance, and service firms—the last illustrated by chain stores, importers/exporters, publishers, and advertising agencies. The executives fell into three groups: family businessmen "whose fathers, grandfathers . . . began the firm they now head"; bureaucrats who "climbed to the board through the management hierarchy"; and entrepreneurs "who began the company they now head, and have built it up over the years or . . . by merger . . . or takeover have put together a large firm." The interviews were held in 1974-76, principally with executives located in London and its vicinity. The schedule of 20 questions, most of several parts, was put to 60 respondents; after that, certain changes in the list involved limitations on final analysis of the data.

The question posed in the study is, "what precisely are the dominant values of our society? If they are to be found among any grouping, then presumably the business elite, above all, will hold them." Be it said that the personages queried expressed their views with becoming modesty. As a sort of backdrop to his inquiry, Fidler summarizes, with evident skepticism, the role of the corporate manager, no longer an owner, as set forth by the Americans Berle, Means, and Galbraith. Particularly, will managers "become responsive to the needs of society and pursue socially desirable aims as well as, or instead of, profit?"

Among Fidler's findings: his top executives were apt to have attended public—in Britain, this means private—preparatory schools and a university. "No evidence for a deviation from the traditional goals of capitalism. Profit remains the aim and measure of success." Moreover, "Top businessmen . . . ascribe to unions a functional role, for managers as well as members . . . a convenience for both groups." A majority do not think that they have particularly high status within the middle class, do not think themselves wealthy, regard the professions or civil service as more esteemed than business, and aspire to no links with the aristocracy.

BROADUS MITCHELL

Rutgers University
New Brunswick
New Jersey

FIELD, JOHN H. *Toward a Programme of Imperial Life: The British Empire at the Turn of the Century*. Pp. xiv, 256. Westport, CT: Greenwood Press, 1982. $29.95.

In this wide-ranging volume John Field sets out to explore the structure of values and attitudes that sustained British imperialism during its heyday in the last decade of Victoria's reign. He is not concerned with the expansion of the empire, nor—apart from a lively first chapter—with the controversies to which that expansion has given rise. The book is instead an excursion onto the treacherous ground of culture and personality studies; its focus is the way a national "commitment to empire" was made to appear as an appropriate and logical extension of central Victorian values, above all the cultivation of a manly character and the disinterested pursuit of duty. Field shows how an array of imperial propagandists endeavored to convince first the educated elite, and then the general public, that a sound character, the nation's destiny, and the spirit of empire interlinked and reinforced each other.

Much of this is of course familiar. As we know from the work of Bernard Semmel, the social Darwinists, obsessed with decline, constantly reiterated such themes, and they suffuse Kipling's work as well. Although Field devotes a chapter to the theorist Benjamin Kidd, his central figures are the lesser-known journalists and publicists who in his view played the major role in "co-opting the masses" for the imperial venture. The heart of the book, and its most sucessful chapters, are the two focused upon George W. Steevens, correspondent of the fledgling *Daily Mail,* who died tragically at 30 during the Boer War in South Africa. Steevens's success, Field points out, derived from his facile stereotyping of alien peoples so that the Englishman's reassuring sense of his own superiority was never called into quetsion, together with his shaping of collectively shared symbols of English self-reliance and enterprise in the service of empire. Of these the most potent was Lord Kitchener at Khartoum, whom Steevens portrayed, Field wrote, as "the cutting edge of a machine-like civilization demonstrating its superiority, as it must to retain it." Field concludes the volume speculatively by delineating what he sees as a distinctive set of character traits that predispose men to support empire, "a sort of *homo imperiosus.*"

Much of Fields's argument remains unproved, for it is by no means clear how far his publicists converted anyone to the cause of empire; indeed given the elusive nature of his analytic categories—what, after all, is character?—his theories may well be unprovable. The book is flawed too by a heavy and rather wooden writing style. Still it remains a provocative introduction to a subject of enduring fascination, which others will no doubt pursue in greater depth.

THOMAS R. METCALF

University of California
Berkeley

UNGER, ARYEH L. *Constitutional Development in the USSR: A Guide to the Soviet Constitutions.* Pp. viii, 310. New York: Pica Press, 1982. Distributed by Universe Books. $30.00.

The present volume is meant "to fill a gap in the textbook literature on Soviet government by enabling students to familiarize themselves with the contents of Soviet constitutions—aided, it is hoped, by commentary and annotation." Unger is senior lecturer in politics at the Hebrew University of Jerusalem.

All four constitutions—1918, 1924, 1936, 1977—are analyzed, with approximately two-thirds of the pages devoted to the Stalin constitution and the one of 1977. Unger finds the 1918 constitution "the most interesting . . . for it still reflected the thrust of the revolution. . . . There were no concessions to the principle of political equality, to inherent freedoms protected by an independent judiciary . . . or to some form of separation of powers."

I can write with little demur that Unger has delivered what he has promised. The texts, translated from the Russian by himself, are there, and the commentary is clear if sometimes terse. He argues convincingly that a fundamental Soviet constitutional doctrine "is the premise that all legal rights of the individual emanate from the state and are enumerated in positive law. There are no natural or inalienable rights of man." From this point of view a key conclusion follows:

> The basic difference between Western constitutionalism, which regards the constitution as the legal embodiment of a mechanism enabling society to control its power holders, and Soviet 'constitutionalism', which regards the constitution as the legal embodiment of a mechanism enabling the power holders to control their society, is as valid today as it was at the beginning of Soviet statehood.

Concerning the most recent constitution, Unger comments, "Probably the most interesting, if still puzzling, fact about the 1977 constitution . . . is its inordinately long period of gestation." And now that Mr. Brezhnev, with whom this constitution was so much associated, is no more, one obviously wonders what constitutional change is in the offing, and when.

Various areas exist for possible criticism or disagreement. In terms of form I would have preferred to have had all authors mentioned in footnotes, such as Timasheff and Hazard, included in the select bibliography. In terms of content it would have been interesting to find a discussion of Western—not just Soviet—constitutional illusion and reality: for example, the constitution of the consulate government of Napoleon Bonaparte.

All in all it is a nicely put-together book, despite some typos and misprints. Libraries and students should find it very useful.

DAVID HECHT

Pace University
New York

YANEY, GEORGE. *The Urge to Mobilize: Agrarian Reform in Russia, 1861-1930.* Pp. 599. Urbana: University of Illinois Press, 1982. $24.95.

In this rather long work Yaney challenges traditional views that both the search for power and reactionary agendas motivated the Russian government's approach to the problem of agrarian reform. On the contrary he finds that each successive wave of reformers, whatever their individual ambitions or ideological orientation, had only the best of intentions when they acted to improve rural life, even though each new set of changes was predicated upon the destruction of the solutions that had been applied by its predecessors.

Yaney seeks to put into perspective the ongoing interaction between the peasantry and state-sponsored reform of rural life during the years 1861-1930 in order to make more understandable the reforms mounted by P. A. Stolypin directly following the 1905 revolution. The phrase "the urge to mobilize" in the title of the work flows from Yaney's perception that reformers in the capital were often poorly informed about the actual state of peasant life, and in any event they sought solutions to the structure of rural life in a priori visions of harmony and order that were written into law and imposed upon the countryside. Yaney sees much of this as well-intended and high-minded rather than self-serving and regressive; the fault often lay, in his view, in the lack of substantial and meaningful information at the disposal of the central government, resulting in efforts to enhance peasant autonomy that often produced results opposite to the reformers' original intentions. Probably no better example of this can be found than Yaney's account of the introduction of the system of land captains in 1889; in this regard he sees the chief architects of the land captain statutes, D. A. Tolstoi and A. D. Pazukhin, as having intended a progressive reform of peasant life at the local level by providing the means for the state to protect the peasants from exploitation. Yaney further alleges that such well-placed and influential reactionaries of the 1880s as K. P. Pobedonostsev, M. N. Katkov, and Prince V. P. Meshcherskii were not calling the shots in the conceptualization of the reform; indeed he finds that they played no substantive role in the evolution of the legislation that resulted in the creation of the land captain system. Yaney fails to demonstrate fully this latter point, however, basing his case largely upon the lack of evidence to the contrary.

While this work is a fascinating and stimulating contribution to the literature, it also is overwritten and at times borders on the idiosyncratic. Yaney makes unusually broad use of the first person singular, injecting himself unduly into the text, and the reader is at times unsure whether he intends a judicious conclusion or is indulging in hyperbole or irony. Extensive lapses into metaphor occasionally obscure rather than illuminate the point he intends to make, sometimes leaving the reader uncertain as to which conclusion to draw. As a consequence the work is likely to be confusing to the novice, while the more knowledgeable reader can both profit from Yaney's prodigious research and lively intellect and find in the work a challenge to long-held views hitherto accepted as conventional wisdom.

FORRESTT A. MILLER

Vanderbilt University
Nashville
Tennessee

UNITED STATES

BLECHMAN, BARRY, ed. *Rethinking the U.S. Strategic Posture.* Pp. xii, 308. Cambridge, MA: Ballinger, 1982. $14.95.

HUNTINGTON, SAMUEL P., ed. *The Strategic Imperative: New Policies for American Security.* Pp. x, 360. Cambridge, MA: Ballinger, 1982. No Price.

There is hardly a question more important now, or more debated, than the proper degree of American military preparedness for the 1980s. The two books reviewed here, offered by a single publisher, very nicely and responsibly illustrate the dimensions of this debate. The volume edited by Samuel Huntington presents a well-stated set of the more hawkish arguments, with most of its chapters' authors anticipating major difficulties for the United States in the Soviet military growth that has taken place since 1970, and in the allegedly too arms-control-minded approaches of American administrations in the same years. The Blechman collection by contrast presents a more middle-of-the-road-to-dovish view, including some contributors who would share the alarm about the Soviet menace, but others who are far less convinced that our military problem is so clear or real.

Both books are responsible and carefully reasoned, and hence never fall into the trap of becoming polarized caricatures of the extremes of the debate. There is in fact some substantial overlap in perspectives, illustrated obviously by the presence of Joseph Nye as an author in both collections, but illustrated equally importantly in the Betts and Friedberg contributions to the Huntington volume, stipulating a number of ambivalences and confusions about the exact nature of the Soviet menace, and about the past debate on U.S. nuclear targeting strategies and options.

Conversely the Hyland and May contributions to the Blechman collection can also hardly be seen as soft-minded on possible Soviet threats, as the team assembled for this Aspen Institute book was designed to be balanced in the perspectives it brought to bear.

The Huntington book, while well researched and well argued, thus comes off as the less balanced of the two, and the more predictable. Criticisms are to be found throughout its pages for all the alleged failings of past liberal perspectives on arms matters, ranging from system analysis to female military service, to arms control to the all-volunteer force. Only one chapter, that by Nye on the energy problem, focuses on a major source of trouble other than the Soviet Union. The Huntington book, the Nye chapter included, is thus generally pessimistic, painting the United States as in a great deal of trouble, with most of the trouble being due to communist power, and communist craving for more power.

In his own chapter Huntington advocates a strategy of conventional retaliation, a mode of retaliation from Western Europe that might be exploited to deter Soviet aggressions elsewhere around the world, which amounts to a counteroffensive to be launched "simultaneously with the enemy's offensive." While carefully enough defined and thought through, this endorsement of yet another strain on preemption and first strike might seem to many Americans exactly the wrong direction to go.

Taken together these two books should provide the student and the general reader with an excellent introduction to the 1980s U.S. strategic debate. As the United States remains convulsed with exaggerated advocacies of the MX, or of the nuclear freeze, the arguments presented here remain on a higher and more scholarly plane.

GEORGE H. QUESTER

University of Maryland
College Park

DUFFY, JOHN J. and H. NICHOLAS MULLER III. *An Anxious Democracy: Aspects of the 1830s.* Pp. xii, 172. Westport, CT: Greenwood Press, 1982. $27.50.

This book explores familiar tensions of Jacksonian America from a very specific perspective. It consists of articles on men and events in northwest Vermont between 1835 and 1839, plus the limited provision of a general historical framework. The key articles focus on a controversial revival conducted by the sensational evangelist, Jedidiah Burchard, Vermont responses and interventions in the *patriote* rebellion across the border in Canada, and the opposition of James Marsh, George Perkins Marsh, and other educated Vermonters to hysterical revivalism and wild campaigns for American military involvement in Canadian affairs.

The articles have value as local history, and Duffy and Muller make a convincing case that men of education and social prominence like the Marshes were elitists, sharing a common conviction that French Canadians were an inferior breed unfit for self-government, and that their Vermont supporters were the same Jacobinical rabble attracted to religious mountebanks like Burchard. Duffy and Muller demonstrate that the pro-*patriote* Vermonters were in fact drawn from a wide cross section of the community, and rightly stress the importance of both the Canadian rebellion and the Buchard revival as mirrors of existing social and political tensions in Vermont. Useful material on the early French Canadian settlers in Burlington and their negative attitudes toward the rebellion also lies buried in the footnotes.

The book breaks little new ground in the treatment of migration and economic depression in postpioneer Vermont. With regard to its key topics it adds only a little to our knowledge, and Duffy and Muller admit that "the various Vermont responses to the 'Patriote War' reflect the larger national responses." Their own footnotes belie the claim that the literature on American-Canadian relations, including the Patriot Hunters movement, is very slim. The Marsh cousins and their circle have also been the subjects of scholarly attention, though Duffy and Muller have some valuable remarks about their social philosophies and intellectual accomplishments.

Other sections of the book are very uneven. Good use is made of primary sources to describe Buchard's revival techniques and the organization of opposition, but recent work on rivivalism at Middlebury College—a key institution—and the valuable study by Paul Johnson of revivalism in the western New York from which Buchard came are ignored. Even more surprising is the absence of any discussion of the roles of women in popular religious movements, a dimension so usefully explored by Nancy Cott, Mary P. Ryan, and others for regions adjacent to Vermont. Finally, the book seems excessively expensive, given that there are only 110 pages of text, and nearly 40 of those have appeared in print elsewhere.

LOUIS BILLINGTON

University of Hull
England

LAKE, ROBERT W. *The New Suburbanites: Race and Housing in the Suburbs.* Pp. xvii, 303. New Brunswick, NJ: Rutgers University, 1981. No price.

During the last decade black suburbanization has outpaced the suburbanization of whites, and like whites most blacks own their houses in the suburbs. Does this movement of blacks to the suburbs presage a change from residential segregation in the central cities to integration in the suburbs?

This book presents carefully marshaled evidence from a survey of 1004 black and white homebuyers, and 20 real estate brokers in six selected suburbs in New Jersey, and concludes that blacks' recent suburbanization offers them less than full realization of the American dream of homeownership as a source of security and equity accumulation.

This study addresses such questions as, What are the institutional and individual barriers to black suburban homeownership? How does racial discrimination add to the cost, time, and difficulty of the black homebuyer's search for housing? What is the impact of such discrimination on housing prices, resale value, and equity accumulation in various suburban residential areas? What is behind black and white attitudes toward residential integration? What is the perspective and role of the real estate agent, the key market intermediary, in regard to the extent to which blacks and whites do or do not become immediate neighbors?

Following are the major findings of the study. The blacks are moving to the older suburbs but remain residentially segregated within these communities. Some blacks, representing the highest-status segment of the black population, are to be found in predominantly white neighborhoods, and in these areas the income of blacks exceeds that of their white neighbors. Those whites who do buy homes in integrated neighborhoods tend to be young, first-time homebuyers, for whom lower housing prices rather than any commitment to integrated housing appears to be the dominant motivation for purchase.

Blacks are more likely than whites to purchase houses in integrated or all-black neighborhoods, to purchase black-owned units regardless of neighborhood racial composition, to use a black real estate agent, and to experience discrimination in the process of searching for and buying their houses. Compared to whites, black housing search is protracted, and this protraction appears to be due to some combination of overt discrimination and black adaptive techniques to avoid such discrimination. Prices for houses decrease for whites as well as for blacks as the neighborhood proportion of black population increases.

When blacks buy they generally pay less than whites do for equivalent housing. When they sell they obtain less for their houses than whites do. The selling markdown for blacks exceeds the purchase markdown, regardless of neighborhood racial composition.

> The price spread is most notable in Black-Entry neighborhoods, where the . . . discount obtained by black buyers is far exceeded by the . . . loss experienced by black sellers. This is reflected in analysis of price changes, where white-occupied units appreciate at a faster rate than black units or units in integrated or black neighborhoods.

In short there are substantial negative consequences for black equity accumulation. Whites, who account for the major share of the suburban housing market, either avoid buying black-owned units or buy heavily marked-down houses in integrated neighborhoods. Overt white avoidance is a major component of market weakness for black-owned units; however, there are institutionalized market mechanisms that arise from and reinforce white preferences. Real estate brokers effectively reduce the level of demand for black-owned houses by steering whites away from such units. Even if black-owned units for sale are included in a multiple listing service, real estate brokers deem MLS information to be confidential and dissemination of such information to prospective buyers to be entirely at the broker's discretion. Furthermore white demand for black-owned units may be additionally curtailed if black sellers list their houses with a black real estate broker who does not have access to a pool of prospective white buyers.

Blacks and whites have unequal access to and use of housing market information. For blacks restricted access to information channels for buyers raises entry costs, and for sellers it reduces the eventual payoff from home ownership. Additionally of course housing discrimination leads to residential segregation.

Robert Lake's study is first rate. It is well written, well organized, and analyzes very cogently the data for the subject matter of this study. One small point may be raised as to the appropriateness of the use in this study of the X^2 statistic when the expected cell frequencies in a number of instances are much too small.

SURINDER K. MEHTA
University of Massachusetts
Amherst

ROSTOW, W. W. *Europe after Stalin: Eisenhower's Three Decisions of March 11, 1953.* Pp. xiii, 206. Austin: University of Texas Press, 1982. $25.00. Paperbound, $8.95.

W. W. Rostow's, *Europe after Stalin,* the third in a series of studies exploring the policymaking process in detail, explores the decisions taken by the Eisenhower administration during the period immediately after the death of Stalin. This excellent narrative, with a highly informative and desirable series of appendices, seeks to relate the interaction of personal predispositions and institutional forces insofar as they help to explain the rationale and timing of decisions taken during this period.

As a study of governmental decision making this effort is highly focused, covering a select number of responses over a relatively short period of time. Specifically, did the death of Stalin represent a historic moment during which progress on German reunification and U.S.-Soviet rapprochement was conceivable? While this question is of course unanswerable, the look provided by Rostow is invaluable in determining the impact of images held by policymakers and their resultant interpretation of events.

The contribution of this study lies in its detailed analysis of the evolution of the eventual Eisenhower response as it was subjected to bureaucratic and individual pressures. As such it affords the reader with an insight into the intricacies of the governmental foreign policy decision-making process rarely provided. At the same time there is a tendency to get lost in the detail and to fail to account for other factors which, if they did not affect the immediate debate, certainly determined the general discourse. In particular Rostow places heavy emphasis on Eisenhower's personal reluctance to negotiate with the Russians as accounting for the delay in the administration's response to Stalin's death. While assuredly accurate, the impression left is one of personal failing, reinforced by the ideological rigidity of his secretary of state, John Foster Dulles.

The operative mode of analysis seems to be one in which personal responses are judged according to how those persons judge, not other persons, but institutional momentum. In this case not enough is mentioned concerning the existing perceptions abounding within U.S. domestic circles concerning the wartime conferences and the so-called monolithic communist movement. Whether for domestic consumption or not, negotiations with the Soviet Union was a topic that transcended the death of Stalin and the question of German reunification. It spoke to the core of the nascent U.S.-Soviet rivalry. And while institutional and personal dynamics may at times enlighten the decision-making process, it is questionable that given the highly ideological charge to the atmosphere of U.S.-Soviet relations at the time—the McCarthy era—reasoned policymaking was feasible.

I do not mean to suggest that Rostow's study is seriously flawed. As an exploration of institutional and personal dynamics it is very welcome. It enlightens our understanding of executive foreign policy decision making, and for that alone it is highly useful. However, in purporting to explore the relationship between ideas and action, it fails to lift its horizon beyond the individual and institutional motivations of policymaking and explore the elite engendered Weltanschanung that fosters them in the first place.

LOUIS FURMANSKI

Blacksburg
Virginia

SOCIOLOGY

AMATO, JOSEPH ANTHONY, II. *Guilt and Gratitude: A Study of the Origins of Contemporary Conscience.* Pp. xvi, 219. Westport, CT: Greenwood Press, 1982. $27.50.

For Amato, the "philosophers' vision makes the world around us," and the "philosophers' dreams are within us." They "are our conscience—and we are their dreams." He thus emphasizes the presumed significance of selected poetic speculations rather than of evidences of popular sentiments—except when it suits his purpose to try to refute Marxist ideas about the influence of social class. In his view the nation-state overshadows class in shaping "the new conscience."

Amato believes that "Guilt and gratitude express the radical difference between the ethics of modern man and traditional man." He is intrigued by the ethical polarities "between a kind of gratitude that requires obedience to a specific established order and a type of guilt that demands the universal realization of human potential." He contends that probing "this division and its ramifications—its hold upon our feelings, sensibilities, assumptions, language, and thus consciousness itself—is to help free conscience from debilitating contradictions that deny clarity to its power to survey and judge human experience." That freedom is needed for "the new conscience" for which he hopes.

Amato claims that "All humanisms, whether they are classical or Christian, elitist, populist, aesthetic, or progressive, have been melted down at their core by revelations of twentieth-century events." He thus sees the undermining of what he calls the "three fundamental assumptions of all humanisms—man's mastery of self, man's mastery of nature, and man's sympathy for man." This is a utopian version of what has long been typical in the aspirations of humanist social scientists. They try to help people intelligently control more of their affairs, relate more constructively to nature, and broaden their social concerns and responsibilities.

Humanisms are confusingly varied and contradictory because of the uses to which the term has been and is being put by propagandists for special interests. Amato is apparently discouraged by the confrontation of humanist ideals by the "ignorance, selfishness, conformities, vindictiveness, and other emotions that motivate modern societies." He even goes so far as to assert, "Sanitation, social welfare, education, suffrage, and all the other tools of progressive reform have not yet taken either the suffering or the demons out of the human condition." But they have made life for millions more livable.

Nevertheless, Amato still concludes, "We must give human measure and direction to the forces that encompass us," an echo of Protagoras's famous humanist dictum. But then he departs from the Protagorean faith in man as "the measure of all things." He contends that rationalism "has been challenged by an entire range of cultural and political expression." He adds, "Against human pain, suffering, and death, progress—and what good men and women can hope for collectively—may not be enough." He therefore turns to "seek not just to honor God, parents, and the dead . . . but to end this terrible strangeness among us—to make humanity more than the abstraction it presently is."

ALFRED McCLUNG LEE

Drew University
Madison
New Jersey

HOLLAND, NORMAN H. *Laughing: A Psychology of Humor*. Pp. 232. Ithaca, NY: Cornell University Press, 1982. $16.95.

In the first half of his book Holland reviews 200 theories of laughter, humor, and comedy. They all founder, he contends, because they neglect the question, Why does one person laugh?

In the rest of his book he seeks redress by asking, Why does Ellen laugh? He interviewed Ellen for an hour. She was asked why she laughed at the cartoons in Kliban's *Never Eat Anything Bigger Than Your Head,* which she had first read three years earlier. He molds her laughter and reflections into a theory: "Ellen laughs," he argues, "when she recreates her identity." The cartoons that make Ellen laugh do so because they confirm a theme in her identity, namely, "to give and so turn the 'other' from a missed to an imagined to a giving humanity." I confess that I do not understand why Ellen laughed. I did not find Kliban's cartoons funny or humane. What is worse, Holland's theory does not help me understand, and it should have.

Holland generalizes: "We laugh when we have recreated our identities though a stimulus suddenly and playfully." Neither the meaning of the terms "identity," "play," and "recreation," nor their relation to one another is made crystal clear. Not all recreations of identity lead to laughter, such as art, sex, eating a madeleine. Not all laughter stems from recreations of identity, such as nervous laughter, often a sign of self-deprecation, or laughter induced by tickling, nitrous oxide, disease. Holland's theory never asks why the playful recreation of identity should lead to laughter: it does not explain laughter.

The relation of laughter to context, society, and culture deserves more attention. So does the relation between comedy and religion; see Auden's article "The Globe" for pertinent remarks. What Holland says about recreation sometimes recalls Tillich's concept of centering in *Systematic Theology*. Centering leads to freedom and joy; lack of it, to bondage and despair. Might not laughter reveal the spirit as well as the mind?

Two of Holland's ideas merit serious attention. First, we should look at individual laughers laughing; and second, laughter is a function of one's personality. Unfortunately the theory's logical and methodological faults deprive it of explanatory power and obscure its insights.

SANFORD B. STEEVER
University of Chicago
Illinois

LERNER, MELVIN J. and SALLY C. LERNER, eds. *The Justice Motive in Social Behavior: Adapting to Times of Scarcity and Change.* Pp. xvii, 494. New York: Plenum Press, 1981. $32.50.

During the early 1970s a number of events dramatically underscored the scarcity of certain vital resources. The most significant attention was paid to the rapid rise in the price of oil. The increased price for this allegedly scarce commodity shook the very foundation of certain Western economies. The complex array of forces that increased the price of oil and other energy sources continue to plague basic industries of the United States. This phenomenon has contributed to the presently high rates of unemployment and accelerated decline of certain areas of the country. Quite clearly the scarcity of certain resources and the socioeconomic context in which this phenomenon operates can have a potentially disruptive and profound impact upon individuals and their communities. As such this is a particularly fertile area of inquiry for social researchers concerned with questions of justice and change.

In *The Justice Motive in Social Behavior: Adapting to Times of Scarcity and Change,* Melvin J. Lerner and Sally C. Lerner attempt to examine the question of social justice within the context of economic contraction. The book is divided into three sections. A broad overview is developed in the first series of chapters. The editors attempt to define both the various facets of justice and a world economy in transition.

The subsequent portion of the analysis focuses on the various procedures associated with justice. Much of this discussion centers on certain theoretical and empirical relationships between children and the justice impulse, macro and micro definitions of justice, sociohistorical factors that shape the justice motive, and retribution as a mode of justice.

The final section examines the changing form of social justice within a number of institutional settings. The authors in this section discuss personal relationships during a period of socioeconomic transition, the distribution of scarce resources in the marketplace, and the justice motive in legal settings. Quite clearly these institutions were selected because they are the fundamental underpinnings of our culture and consequently provide a forum for identifying and resolving justice disputes. Concomitantly these institutions are on the cutting edge of certain kinds of social change. They therefore frequently forecast shifts in perspective or policy that will affect the larger social order.

This volume makes a substantial contribution to the emergent interdisciplinary literature on social behavior during a period of scarcity. The editors have carefully selected a series of articles that contribute to the reader's greater understanding of the dynamic interaction between the individual, collective, and various definitions of social justice.

Melvin Lerner and Sally Lerner's brief introductions to each section of the book are thoughtful and clear. They identify those analytic threads that make the chapters cohere. They also discuss certain explicit and subtle ties between entire sections of the book. Finally the authors frequently raise pertinent questions that highlight critical facets of a reading.

Despite the many strengths of this book, certain weaknesses detracted from its overall quality. Too frequently terms such as "equity" or "distributive justice," although defined early, were redefined in great detail in subsequent chapters. Despite the occasional need to do this because the terms were being discussed within new contexts, the frequency contributed to redundancy of certain portions. Additionally the promise outlined in the initial section was not consistently fulfilled in subsequent chapters. Too frequently insufficient attention was paid to the relationship between scarcity and justice. The overriding question of resource scarcity was too often lost in definitional discussions of justice and micro experiments in social psychology. In fairness, it should be said that this imbalance was addressed in the final section of the book. Finally many of the authors' definitions of problem and solution to the dilemma of maintaining justice in our increasingly impoverished world are tied to static definitions of reality. Too often the authors' foundations for change are the marketplace and legal institutions.

Despite this book's deficits, its effort to grapple with the meaning of justice during a period of dramatic socioeconomic change is most impressive. Many of its shortcomings can be attributed to the authors' risky struggle with material that has rarely been treated by other academics. They open critical lines of inquiry that one hopes will stimulate additional research on the interaction between resource scarcity and social justice at the level of the individual, community, and nation-state.

In sum this is a pioneering work that should prove quite valuable to professionals in the fields of social psychology, criminal justice, social policy, and the law.

MICHAEL FABRICANT

Hunter College
New York

REAMER, FREDERIC G. *Ethical Dilemmas in Social Service*. Pp. xiii, 280. New York: Columbia University Press, 1982. $17.50.

In this book Frederic G. Reamer, professor of social work, addresses a number of questions concerning the methods used to justify social work principles and values. His premise is that fundamental ethical questions in social work derive from ethics and moral philosophy, and concern issues of metaethics and normative ethics. He finds the Code of Ethics of the National Association of Social Workers (NASW) to be deficient in guidelines that can help resolve differences of opinion among practitioners. The specific need is for objective ethical standards to reconcile value conflicts. Reamer believes the challenge is to develop guidelines for ethical analysis that will help social workers identify ethical issues and reason about the ethical dilemmas they encounter in their practice. These dilemmas encompass questions of right and wrong and focus on the professional ethics of the practitioner, not on the morality of the social service recipient's behavior.

Presented for consideration are Alan Gewirth's argument that ethical principles can be justified. Inferences from Gewirth's conclusions offer guidelines that Reamer suggests can be applied by social workers to resolve conflicts of value and duty in an age of increasing complexity. Most helpful are the case examples Reamer provides to assist the practitioner—or teacher and student, or grievance arbitrator—in the direct application of the six presented guidelines to actual practice situations involving ethical dilemmas in services to individuals and families, in social planning and policy, and in relationships among practitioners.

This book has implications for the practice of social work. Reamer's rational approach to the resolution of ethical dilemmas may be viewed by some who prefer situation ethics as too prescriptive. His intent is to have social service workers advance from intuitive practice wisdom toward greater reliance on a logically consistent set of norms as practice guidelines. Such a practice goal carries implications for professional degree curricula and continuing education.

By suggesting that ethical conclusions emanating from the NASW Code of Ethics tend to be ambiguous and at times contradictory, Reamer raises an urgent question for the social work profession to address: do ethical principles, which appear to be largely based on professional opinion, "have sufficient validity to compel social workers to act in accord with them"? Ethical issues in social work—such as those having to do with rights to self-determination, freedom, and well-being, or the role of the state, or the distribution of services and limited resources—certainly are timely discussion subjects and require more than a utilitarian approach. If congruence with specified values is to be achieved, a rational approach must be made to operate in contemporary social work practice.

I find Reamer scholarly in his examination of ethical dilemmas in social service. His technical competence in philosophical ethics must be acknowledged. He has made a noteworthy contribution to the professional literature with his book.

FLORENCE P. ENGELHARDT
Arizona State University
Tempe

ECONOMICS

MEAGHER, ROBERT F. *An International Distribution of Wealth and Power: A Study of the Charter of Economic Rights and Duties of States.* Pp. xvii, 303. Elmsford, NY: Pergamon Press, 1979. $37.00. Paperbound, $15.00.

This book is an enduringly useful work. There is much that it is not, but within its chosen parameters it is a careful, thorough, and level-voiced examination of the Charter of Economic Rights and Duties of States. The charter was approved by the General Assembly of the United Nations in 1974, during the aftermath of the oil price increases induced by the Organization of Petroleum Exporting Countries (OPEC), when Third World hopes for a new international economic order ran high. The charter is an expression of those hopes.

The study is organized chronologically. It provides historical background on the various international discussions of these issues during the preceding decades. It then offers a very detailed account of the charter itself, providing us, article by article, with information not only on content but on the complex back-and-forth that took place during the deliberations of the General Assembly. This is followed by a detailed discussion of the multitude of international discussions in the three years subsequent to the adoption of the charter.

For the most part the study is informative rather than analytic. It is not written by an economist and does not attempt to assess the economic significance of these events for the development of Third World nations. Similarly no attempt is made to analyze these events within the larger framework of the foreign policy objectives of the key players. For instance there is little discussion of how this coming together of Third World nations was viewed by American policymakers, what their objectives and tactics were, and how their actions were conditioned by larger foreign policy objectives.

The book is not written for the general reader; it will be of greatest value to those already involved in these issues, perhaps engaged in research on their own and desiring to know exactly what happened and when. At the same time, and as a result of its detailed chronological approach, the book succeeds in demonstrating, almost without trying, certain very fundamental truths: that the call for a New International Economic Order is not something that appeared spontaneously in the wake of OPEC; that the various objectives and conceptions of rights and duties of states each have a slowly evolving history; that over the long haul there is considerable change in the basic terms of economic interaction of nation states; and that the call for a New International Economic Order is best seen as a phase in that longer-term evolution. It has the distinct value of causing the reader to lengthen the implicit time frame used in perceiving the current moment in relations with the Third World.

JEROME SEGAL

Washington, D.C.

MUNNELL, ALICIA H. *The Economics of Private Pensions*. Pp. xiii, 240. Washington, DC: Brookings Institution, 1982. $22.95.

Newspaper stories on the plight of the social security system abound, but little is ever written about the economic viability of private pensions. Alicia Munnell remedies this situation with her splendid book, which is a sweeping analysis and critique of private pension plans and how they are related to social security, the tax structure, and, among other things, capital markets. While many economists rightly focus on the dubious future of the social security system, Munnell argues that equal attention should be paid to the instability and underfunding of private pension plans. She carefully reviews the impact of the Employee Retirement Income Security Act of 1974, which required that private pension plans set aside a certain percentage of their funds in vested accounts. Despite this legislation she examines how corporate bankruptcies—for example, Chrysler, International Harvester—could have imposed severe financial commitments on an already strained federal pension system.

The book examines pension plans in unionized industries, usually concentrated in higher-than-average paying jobs. It gives a detailed analysis of the extent to which managers of pension funds have erred by investing in low-yielding securities instead of higher-yielding investments. Munnel concludes her book with a warning that unless some form of indexing occurs in private pension plans, they are likely to be less popular in the future.

Munnell's analysis is weak in one respect. She does not fully delineate the basic inequities in compensation between private and social security systems. For high-income earners a failure in a particular private pension fund would not produce as much economic hardship as low-income earners would experience if social security payments were reduced or if the system collapsed. Thus, while some private pension funds remain seriously underfunded, their failures may not be as harmful as the failure of the social security system.

Despite this flaw this book represents a comprehensive and thoughtful analysis of private pension funds. It is a welcome addition in an area where rhetoric has replaced substance all too often. For those interested in the viability of private pension plans, this book is an excellent place to start.

LOU FERLEGER

University of Massachusetts
Boston

OUCHI, WILLIAM G. *Theory Z: How American Business Can Meet the Japanese Challenge.* Pp. xii, 244. New York: Avon, 1982. $2.95. (Originally published by Addison-Wesley, 1981. $12.95.)

In the wake of intensified international competition, the Ouchi volume has played an important function in drawing attention to Japanese and American management systems in an era of self-examination. Basically the book outlines the adjustments that could be undertaken by American firms to improve productivity and the quality of worklife. The essential ingredient for this transformation involves a drastic alteration in employee relations.

Theory Z has been a highly successful marketing title that refers to companies with features similar to Japanese firms. The Theory Z company offers trust as the key element to encourage worker concern for quality and productivity. Trust is demonstrated through the exceptional responsibility shown for the welfare of workers. Lifelong employment encourages loyalty and commitment among employees. The long run with a substantial investment in training is the time framework for both the workers and the company. Promotions are made from within the organization, usually after 10 years of service. Cooperation among employees and managers is induced through an elaborate system of job rotation in which employees learn each aspect of the organization. Cooperation and effective policy implementation are assured through consensus decision making, which can involve as many as 60 to 80 individuals. An equalitarian atmosphere in Theory Z is aimed at avoiding the alienation that results from corporate bureaucracy. A holistic relationship and intimacy in industrial life stresses mutiple bonds of workers to the organization in contrast to the partial relationship that most Western firms have with their employees.

Ouchi suggests that the Theory Z format is not culture-bound to Japan, but will work exceedingly well in Western companies. He provides a series of steps that will lead a Type A company—typical U.S. firm—into a Type Z organization. The objective is to achieve the commitment of employees to the development of a less selfish, more cooperative approach to work. The change should begin with a consistent organizational philosophy and the backing of top management. The more decisions that are shared, the greater will be the benefits of cooperation. Communication is an essential element in creating trust. Surveys and questionnaires will be utilized to facilitate feedback, with the development of interpersonal skills as a top priority for the organization. Unions will have to be integrated into the process. Ouchi does not blame unions for current productivity problems in the United States, but suggests that "the union has been forced into an adversarial position by an unfair and irresponsible management." A number of case histories are provided in the book to illustrate the process. Hewlett-Packard, the California electronics manufacturer, is the best example of a Type Z firm in the United States. IBM, Kodak, and Procter and Gamble come close, while Intel, Eli Lilly, Dayton-Hudson, and even General Motors have adopted some elements of the Type Z format.

While Ouchi maintains that Japan's edge is due not to outmoded plants in the United States but rather to a failure in human relations, we continue to see evidence that the lack of investment in plants and equipment is a problem for American firms. In Nissan's highly automated Datsun assembly plant, the production rate is seven times that of comparable American plants.

The competitive edge of Japan may also involve conditions that would be objectionable in America. Some of Japan's competitive edge may hinge on a degree of exploitation; there is no pension or social security. The educational system is strenuous and categorizes children at four years of age, based on scores on an entrance examination for a special kindergarten; Ouchi saw, however, no negative effects.

Essentially the structure of corporate Japan is devised to benefit male employees of the major firms. Although somewhat vague on the negative impact of some of these policies, Ouchi recognizes that "This structure distributes social costs in a way quite different from the United States." Thus he suggests that we should adapt only those aspects of Japanese management that suit our needs.

One question remaining in the future is whether Japan will ultimately outstrip the United States in advanced technology. The Japanese have placed an increasing emphasis on research and development, but in the past they received most of their basic ideas and technology from the West. Individuality and initiative have been a product of Western culture. Will innovations be forthcoming in a setting where "nothing of consequence occurs as a result of individual effort?"

There is no doubt that *Theory Z* is an exceedingly significant work in a period of catharsis for American management. Few would disagree that the neglect of human relations has characterized corporate organizations in the United States.

RUSSELL BELLICO

Westfield State College
Westfield
Massachusetts

OTHER BOOKS

ATWATER, EASTWOOD. *Adolescence.* Pp. 415. Englewood Cliffs, NJ: Prentice-Hall, 1983. Paperbound, no price.

BEHLMER, GEORGE K. *Child Abuse and Moral Reform in England, 1870-1908.* Pp. viii, 320. Stanford, CA: Sanford University Press, 1982. $30.00.

BERTSCH, GARY K., ed. *Global Policy Studies.* SFE, vol. 53. Pp. 232. Beverly Hills, CA: Sage Publications, 1982. $20.00. Paperbound, $9.95.

BLAUSTEIN, SAUL J. *Unemployment Insurance, Fund Insolvency and Debt in Michigan.* Pp. xiii, 82. Kalamazoo, MI: W. E. Upjohn Institute for Employment Research, 1982. No price.

BOTTIGHEIMER, KARL S. *Ireland and the Irish: A Short History.* Pp. ix, 301. New York: Columbia University Press, 1982. $19.95.

BRAESTRUP, PETER. *Big Story.* Pp. xvi, 613. New Haven, CT: Yale University Press, 1983. $25.00. Paperbound, $9.95.

BROADHURST, ARLENE IDOL, ed. *The Future of European Alliance Systems: NATO and the Warsaw Pact.* Pp. xx, 316. Boulder, CO: Westview Press, 1982, $22.50.

BURKE, EDMUND. *A Vindication of Natural Society.* Pp. vii, 105. Indianapolis, IN: Liberty Classics, 1982. $8.50. Paperbound, $4.50.

BYKOV, O., V. RAZMEROV, and D. TOMASHEVSKY. *The Priorities of Soviet Foreign Policy Today.* Pp. 214. Moscow: Progress, 1982. Distributed by Imported Publications, Chicago, IL. $5.00.

CAMPBELL, JOHN P., RICHARD L. DAFT, and CHARLES L. HULIN. *What to Study: Generating and Developing Research Questions.* SOIM, vol. 6. Pp. 168. Beverly Hills, CA: Sage Publications, 1982. $17.95. Paperbound, $7.95.

CARLTON, DAVID and CARLO SCHAERF, eds. *The Arms Race in the 1980s.* Pp. xxi, 338. New York: St. Martin's Press, 1982. $27.50.

CLARKE, JAMES W. *American Assassins: The Darker Side of Politics.* Pp. xii, 319. Princeton, NJ: Princeton University Press. $18.50.

COCHRAN, CLARKE E. et al. *American Public Policy: An Introduction.* Pp. xix, 436. New York: St. Martin's Press, 1982. $16.95.

COFFEY-LEWIS, LOU. *Be Restored to Health: How to Manage Stress, Heal Your Self, and Be Whole Again.* Pp. 222. Toronto, Canada: Personal Library, Publishers, 1982. $11.95.

COOPER, MARK N. *The Transformation of Egypt.* Pp. 278. Baltimore, MD: Johns Hopkins University Press, 1982. $22.50.

COTTINGHAM, CLEMENT, ed. *Race, Poverty, and the Urban Underclass.* Pp. vii, 208. Lexington, MA: Lexington Books, 1982. $24.95.

CRABB, CECIL V., Jr. *The Doctrines of American Foreign Policy: Their Meaning, Role, and Future.* Pp. xii, 446. Baton Rouge, LO: Louisiana State University Press, 1982. No price.

DAHRENDORF, RALF, ed. *Europe's Economy in Crisis.* Pp. xii, 274. New York: Holmes & Meier, 1982. $24.95.

DAVIS, DAVID HOWARD. *Energy Politics.* 3rd ed. Pp. vii, 323. New York: St. Martin's Press, 1982. $15.95. Paperbound, $8.95.

DAWISHA, ADEED and KAREN DAWISHA, eds. *The Soviet Union in the Middle East: Policies and Perspectives.* Pp. 172. New York: Holmes & Meier, 1982. No price.

DOMINGUEZ, JORGE I., ed. *Cuba: Internal and International Affairs.* Pp. 230. Beverly Hills, CA: Sage Publications, Inc. $22.00. Paperbound, $10.95.

DUDLEY, BILLY. *An Introduction to Nigerian Government and Politics.* Pp. ix, 367. Bloomington, IN: Indiana University Press, 1982. $25.00. Paperbound, $9.95.

DUNN, WILLIAM N., ed. *Values, Ethics, and the Practice of Policy Analysis.* Pp. ix, 241. Lexington, MA: Lexington Books, 1983. $27.95.

DYE, THOMAS R. *Who's Running America?: The Reagan Years.* 3rd ed. Pp. xiii, 285. Englewood Cliffs, NJ: Prentice-Hall, 1983. Paperbound, $12.95.

EICHHORN, W. et al., eds. *Economic Theory of Natural Resources.* Pp. 645. Würzburg, Germany: Physica-Verlag, 1982. Paperbound, no price.

FILLER, LOUIS. *A Dictionary of American Social Change.* Pp. 256. Malabar, FL: Robert E. Krieger Publishing Co., Inc., 1982. $16.50. Paperbound, $10.50.

FINSTERBUSCH, KURT, ed. *Social Problems 82/83.* Pp. 241. Guilford, CT: Dushkin Publishing Group, Inc., 1982. Paperbound, no price.

FLEISHMAN, JOEL L., ed. *The Future of American Political Parties: The Challenge of Governance.* Pp. viii, 174. Englewood Cliffs, NJ: Prentice-Hall, 1982. $14.95. Paperbound, $6.95.

FRUCHT, RICHARD C. *Dunarea Noastra Romania, The Great Powers, and the Danube Question 1914-1921.* Pp. ix, 216. New York: Columbia University Press, distributor for East European Quarterly, 1982. $22.50.

GAINES, HAROLD, ed. *Urban Society.* 2nd ed. Pp. 210. Guilford, CT: Dushkin Publishing Group, Inc., 1982. Paperbound, no price.

GLENNON, JOHN P. and NEAL H. PETERSEN, eds. *Foreign Relations of the United States, 1952-1954.* Vol. XIII, Part 1. Pp. xxvii, 1409. Washington, DC: Government Printing Office, 1982. No price.

GLENNON, JOHN P. and NEAL PETERSEN, eds. *Foreign Relations of the United States, 1952-1954.* Vol. XIII, Part 2. Pp. xxiii, 1088. Washington, DC: Government Printing Office, 1982. No price.

GRABER, DORIS A., ed. *The President and the Public.* Pp. x, 310. Philadelphia, PA: Institute for the Study of Human Issues, 1982. No price.

GREANIAS, GEORGE C. and DUANE WINDSOR. *The Foreign Corrupt Practices Act: Anatomy of a Statute.* Pp. ix, 187. Lexington, MA: Lexington Books, 1982. No price.

HALE, DENNIS, ed. *The United States Congress. Proceedings of the Thomas P. O'Neill, Jr., Symposium.* Pg. xxvii, 360. Leominster, MA: Eusey Press, 1982. Paperbound, no price.

HAYS, STEVEN W. and RICHARD C. KEARNEY, eds. *Public Personnel Administration: Problems and Prospects.* Pp. xii, 372. Englewood Cliffs, NJ: Prentice-Hall, 1983. Paperbound, $15.95.

HOLSTI, K. J. *International Politics: A Framework for Analysis.* 4th ed. Pp. xii, 478. Englewood Cliffs, NJ: Prentice-Hall, 1983. $20.95.

HORVAT, BRANKO. *The Political Economy of Socialism: A Marxist Social Theory.* Pp. xx, 671. Armonk, NY: M. E. Sharpe, Inc., 1982. $35.00.

HOUSE, ERNEST R. et al., eds. *Evaluation Studies: Review Annual.* Vol. 7. Pp. 752. Beverly Hills, CA: Sage, 1982. $37.50.

HUCKFELDT, ROBERT R., C. W. KOHFELD and THOMAS W. LIKENS. *Dynamic Modeling: An Introduction.* Pp. 96. Beverly Hills, CA: Sage Publications, 1982. Paperbound, $4.50.

HUMPHREY, FREDERICK G. *Marital Therapy.* Pp. vii, 248. Englewood Cliffs, NJ: Prentice-Hall, 1983. $19.95.

HUNTER, JOHN E., FRANK L. SCHMIDT, and GREGG B. JACKSON. *Meta-Analysis: Cumulating Research Findings Across Studies.* SOIM, vol. 4. Pp. 176. Beverly Hills, CA: Sage Publications, 1982. $17.95. Paperbound, $7.95.

JANICS, KALMAN. *War and Society in East Central Europe: The Effects of World War II.* Vol. IX. Pp. 240. New York: Columbia University Press, 1982. $25.00.

JOHNSTONE, RONALD L. *Religion in Society: A Sociology of Religion, 2nd ed.*

Pp. viii, 312. Englewood Cliffs, NJ: Prentice Hall, Inc., 1983. $21.95.

JÖNSSON, CHRISTER, ed. *Cognitive Dynamics and International Politics.* Pp. vii, 210. New York: St. Martin's Press, 1982. $27.50.

KEGLEY, CHARLES, Jr. and PAT McGOWAN, eds. *Foreign Policy USA/USSR.* Vol. 7. Pp. 320. Beverly Hills, CA: Sage Publications, 1982. $25.00. Paperbound, $12.50.

KINCAID, JOHN, ed. *Political Culture, Public Policy and the American States.* Pp. viii, 235. Philadelphia, PA: Ishi Publications, 1982. No price.

KIRK, RUSSELL. *Reclaiming a Patrimony.* Pp. 116. Washington, DC: Heritage Foundation, 1982. No price.

KLEPPNER, PAUL et al. *The Evolution of American Electoral Systems.* Pp. xiii, 279. Westport, CT: Greenwood Press, 1981. $29.95.

KNOKE, DAVID and JAMES H. KUKLINSKI. *Network Analysis.* Pp. 96. Beverly Hills, CA: Sage Publications, 1982. Paperbound, $4.50.

KONOPKA, GISELA. *Social Group Work: A Helping Process.* 3rd ed. Pp xix, 236. Englewood Cliffs, NJ: Prentice-Hall, 1983. $19.95.

KRASNOW, ERWIN G., LAWRENCE D. LONGLEY, and HERBERT A. TERRY. *The Politics of Broadcast Regulation.* 3rd ed. Pp. xiv, 304. New York: St. Martin's Press, 1982. $16.95. Paperbound, $8.95.

LABRIE, ROGER P. et al. *U.S. Arms Sales Policy: Background and Issues.* Pp. 88. Washington, DC: American Enterprise Institute for Public Policy Research, 1982. Paperbound, no price.

LaCAPRA, DOMINICK and STEVEN L. KAPLAN, eds. *Modern European Intellectual History: Reappraisals and New Perspectives.* Pp. 317. Ithaca, NY: Cornell University Press, 1982. $29.50.

LEIBZON, B. M. *Unity, Solidarity, Internationalism.* Pp. 376. Moscow: Progress, 1982. Distributed by Imported Publications, Chicago, IL. $5.00.

LIM, GILL C., ed. *Regional Planning: Evolution, Crisis and Prospects.* Pp. viii, 189. Totowa, NJ: Allanheld Osmun & Co., 1983. $15.95.

LINK, ARTHUR S., ed. *The Papers of Woodrow Wilson.* Vol. 40 *November 20, 1916-January 23, 1917.* Pp. xxi, 598. Princeton, NJ: Princeton University Press, 1982. $30.00.

MAGHROORI, RAY and BENNETT RAMBERG, eds. *Globalism versus Realism: International Relations' Third Debate.* Pp. vii, 237. Boulder CO: Westview Press, 1982. $25.00. Paperbound, $12.00.

MAOZ, ZEEV. *Paths to Conflict: International Dispute Initiation, 1816-1976.* Pp. xiv, 273. Boulder, CO: Westview Press, 1982. Paperbound, $21.00.

MASON, ALPHEUS THOMAS, WILLIAM M. BEANEY, and DONALD GRIER STEPHENSON, Jr., *American Constitutional Law: Introductory Essays and Selected Cases.* 7th ed. Pp. xvi, 684. Englewood Cliffs, NJ: Prentice-Hall, 1983. $27.95.

MULLER, ROBERT. *New Genesis: Shaping of Global Spirituality.* Pp. xvii, 192. New York: Doubleday, 1982. $14.95.

PARNES, HERBERT S. *Unemployment Experience of Individuals over a Decade: Variations by Sex, Race and Age.* Pp. ix, 99. Kalamazoo, MI: W. E. Upjohn Institute for Employment Research, 1982. No price.

PATERSON, THOMAS G., J. GARRY CLIFFORD, and KENNETH J. HAGAN. *American Foreign Policy: A History/since 1900.* 2nd ed. Pp. xii, 661. Lexington, MA: D. C. Heath and Co., 1983. Paperbound, no price.

PATTI, ARCHIMEDES L. A. *Why Viet Nam?: Prelude to America's Albatross.* Pp. xx, 612. Berkeley, CA: University of California Press, 1982. $19.50. Paperbound, $10.95.

PATTI, RINO J. *Social Welfare Administration: Managing Social Programs in a Developmental Context*. Pp. xiii, 258. Englewood Cliffs, NJ: Prentice-Hall, 1983. $19.95.

PERRY, RONALD W. *The Socal Psychology of Civil Defense*. Pp. xi, 127. Lexington, MA: Lexington Books, 1982. No price.

PRICE, KENT A., ed. *Regional Conflict and National Policy*. Pp. xviii, 142. Washington, DC: Resources for the Future, Inc., 1982. $18.00. Paperbound, $6.95.

RA'ANAN, URI, ROBERT L. PFALTZGRAFF, Jr., and GEOFFREY KEMP, eds. *Projection of Power: Perspectives, Perceptions, and Problems*. Pp. 351. Hamden, CT: Shoe String Press, Inc., 1982. $32.50.

RIST, RAY C., ed. *Policy Studies Review Annual*. Vol. 6. Pp. 775. Beverly Hills, CA: Sage Publications, 1982. No price.

ROSOF, PATRICIA J. F., WILLIAM ZEISEL, and JEAN B. QUANDT, eds. *The Military and Society: Reviews of Recent Research*. Trends in History, vol. 2, no. 2. Pp. 120. New York: Harworth Press, 1982. $20.00.

SERAFICA, FELICISIMA C., ed. *Social Cognitive Development in Context*. Pp. xi, 283. New York: Guilford Press, 1982. $22.50.

SHORE, BRADD. *Sala'Ilua: A Samoan Mystery*. Pp. xvii, 338. New York: Columbia University Press, 1982. $28.00. Paperbound, $14.00.

SHORT, JOHN R. *An Introduction to Political Geography*. Pp. x, 193. Boston, MA: Routledge & Kegan Paul of America Ltd., 1982. $29.95.

STEVENS, JOHN D. *Shaping the First Amendment: The Development of Free Expression*. Pp. 157. Beverly Hills, CA: Sage Publications, 1982. $15.00. Paperbound, $7.95.

STOESSINGER, JOHN G. *Why Nations Go to War*. 3rd ed. Pp. xiii, 226. New York: St. Martin's Press, 1982. $12.95. Paperbound, $6.95.

STROMAN, DUANE F. *The Awakening Minorities: The Physically Handicapped*. Pp. xiii, 254. Washington, DC: University Press of America, 1982. $23.50. Paperbound, $11.50.

TRIPODI, TONY. *Evaluative Research for Social Workers*. Pp. x, 180. Englewood Cliffs, NJ: Prentice-Hall, 1983. Paperbound, $12.95.

WILLIAMS, THOMAS RHYS. *Socialization*. Pp. xx, 476. Englewood Cliffs, NJ: Prentice-Hall, 1983. $20.95.

INDEX